The Pipe Companion

The Pipe Companion

A Connoisseur's Guide
by David Wright

RUNNING PRESS
PHILADELPHIA · LONDON

(Title page) Courtesy of UST, Inc.
Figural clay-pipe bowl made by Gambier of Paris, c.1850.
The bowl is modeled after the Duchesse de la Valliere,
mistress of Louis XIV (1638–1715).

© 2000 Running Press

All rights reserved under the Pan-American
and International Copyright Conventions
Printed in China

*This book may not be reproduced in whole or in part, in any form or
by any means, electronic or mechanical, including photocopying, recording,
or by any information storage and retrieval system now known or
hereafter invented, without written permission from the publisher.*

9   8   7   6   5   4   3   2   1
Digit on the right indicates the number of this printing

Library of Congress Cataloging-in-Publication Number 99-75091
ISBN 0-7624-0323-3

Designed by Frances J. Soo Ping Chow
Edited by Marc E. Frey and Susan K. Hom
Typography: ITC Berkeley, and Linoscript

This book may be ordered by mail from the publisher.
Please include $2.50 for postage and handling.
But try your bookstore first!

Running Press Book Publishers
125 South Twenty-second Street
Philadelphia, Pennsylvania 19103-4399

Visit us on the web!
www.runningpress.com

Table of Contents

Acknowledgments 6

Introduction 8

Part I: A Brief History of the Pipe..10
    The Origins of the First Pipe 11
    Types of Pipes..14

Part II: The Art of Pipe Making..32
    The Materials 33
    Selecting a Pipe 44
    Smoking a Pipe 50

Part III: The Pipe Makers Directory 56
    Canada...59
    Corsica 64
    Denmark..67
    France..82
    Germany 84
    Great Britain..91
    Ireland...100
    Italy..102
    Japan..120
    Spain..125
    Sweden 128
    Turkey..131
    United States...136

Glossary 188

Suggested Readings 193

Pipe Makers Contact Information 196

Notes..206

Credits..208

Acknowledgements

My experiences working on this book prove again that people in the tobacco business are truly people with character, ability, talent, and sincerity. The people I have met were more than eager to help and share information, contacts, ideas, suggestions, and resource material on a subject held very, very dear to them. I am constantly amazed by the immediate brotherhood, camaraderie, partnership, and bonding I've witnessed. To both new and old friends, I dedicate this book.

As with most books, there are a few key people without whose assistance or guidance this book would never have become a reality. And so it is with me. I owe special thanks to the following people: Frank Burla, antique pipe collector; Benjamin Rapaport, author, scholar, rare book dealer, and antique pipe collector; and Keith Moore, Bill Nunnelly, and Stacy Harmon, all of the Uptown's Smoke Shop.

I also wish to thank: Tom Juranitsch, Lioncrest Inc.; Joe Rowe and Frank Blews, Lane Tobacco; Rob Siegel, The Marble Arch, Ltd.; Alan Schwartz, *Smoke* magazine; Thomas Cristiano, Cristom Pipes; Melissa King, Mastro de Paja; Mark Stewart, pipe smoker and collector; Rex Poggenpohl, pipe enthusiast and collector; Gerard Ezvan, F & K Cigar Co.; David Field, Ashton;

Robert Hamlin, Castello Pipes; Bob Ysidron, Savinelli Pipes; Tim Guillot, Tinderbox; Ira LaPedies, Gatlin-Burlier, Inc; Barney Suzuki, pipe collector, author, and president of the Pipe Club of Japan; Philip Thompson, CAO International; Bill Feuerbach III, S.M. Frank & Co., Inc.; Steve Monjure, Monjure International USA; Beth and Samil Sermet, SMS Meerschaum; and Bill Healey, George Brissie, and Todd Smith of Uptown's Smoke Shop.

Finally, I extend a special thank you to the pipe makers: Shizuo Arita, Alfred Baier, Paolo Becker, J.M. Boswell and his wife Gail, John Calich, J.T. Cooke, Jody Davis, Lee Erck, Ron Fairchild, Per Hansen, Ulf Noltensmeier, Jun'ichiro Higuchi, David Jones, Ed Andrew Jurkiewicz, Lynne Kirsten and Eugene Kirsten, Samuel Learned, Lucille Ledone, James Margroum, Andrew Marks, Clarence Mickles, Elliott Nachwalter, Joan Saladich y Garriga, Denny Souers, Trever Talbert, Mark Tinsky, Julius Vesz, Roy Roger Webb, Steve Weiner, Tim West, and Randy Wiley.

Introduction

As we enter a new millennium, we often notice how our lives change, our country and its values change, and, obviously, technology changes. If we look more closely, however, we will also find elements of continuity. Pipe making is one of these elements of continuity that has witnessed little or no technological advancement. An art form established centuries ago, many of the styles, processes, and techniques found in crafting pipes today date from the earliest days of pipe making.

Today, pipes and pipe smoking are experiencing a resurgence in the United States and abroad. Many of the pipe makers featured in this book cannot keep up with demands.

As a result, pipe enthusiasts are finding themselves looking at many wonderful styles of pipes and even more bountiful blends of refreshing and aromatic pipe tobaccos. The pipe has returned!

A well-crafted pipe is a piece of art to be appreciated, admired, and enjoyed. As with antique cars and fine furniture, good designs, pleasing lines, and fine craftsmanship remain in style for years and command respect and recognition. It is this same standard of quality and beauty found in well-crafted pipes that smokers yearn for, and newcomers to the pipe will quickly sense and recognize.

The following pages highlight carvers and manufacturers of pipes from around the world. Included in the book are both large manufacturing houses and one-person shops or cottage-industry artisans. Inside you'll find a variety of pipes being crafted today. You'll also get an in-depth look at pipe makers, their design philosophies, and the raw materials they use.

This companion showcases pipes from the low end to the high end. Not all finely made pipes have to be expensive. There are many wonderful pipes available that are executed with precision and beautifully styled at very reasonable rates.

So lean back, fire up your pipe, and relax as you explore the world of the pipe.

Part I:
A Brief History of the Pipe

Have you ever wondered how the pipe was invented in the first place, or how the shape, design, and function might have changed from pipe making's infancy to the myriad examples we have today of classical, freehand, and contemporary models?

Clay pipe from Mexico, c.1500

The Origins of the First Pipe

In Eastern North America, the earliest pipes date from the middle Woodland period—about 500 B.C. to 500 A.D. Shaped like a tube or an hourglass, many of these early stone pipes are known as tube pipes. It is believed that these pipes were used both by Shamans to remove illnesses and evil spirits from people and to smoke tobacco. The end of the Woodland period saw the introduction of the platform pipe or monitor-style pipe. Platform pipes have a flat base crowned with a bowl often carved in the form of a human or animal.

During the Mississippian period (A.D. 900 to 1600), the style of pipes changed as they evolved into pieces of sculpture representing animals and humans. Ducks, wolves, bears, human heads and legs, as well as humans performing daily tasks were well represented on stone pipes.

Native American Pipes

Tobacco was treated with great reverence among the Native Americans of this period, and many of their ceremonies and rituals included some form of pipe smoking. In some tribes, in fact, the use of tobacco was restricted to ceremonies. The Natchez Indians, for example, gave their sacrificial victims a large pellet

Pre-historic Iroquois Indians clay pipes, Northeast America.

of tobacco to stupefy them before death. The Creeks and Choctaws offered water, food, and tobacco to travelers while war parties of Creeks traveled for days subsisting only on water and tobacco. Tobacco was their stimulate to help them overcome fatigue during a 50 to 100 mile raid.

Perhaps the most identifiable pipe from the Native Americans of this period is the peace pipe. Historically, the peace pipe served as a passport to traveling visitors; was used to "seal" a trade or some other business venture, or served as a symbol for declaring war or peace. This pipe has become well known through its depiction in television and movies, and its use in Native American pow-wows. The stem of a peace pipe—considered a bridge to the sky or spirits—is its most important part. It was decorated with feathers, cut groves, and quillwork. It is the bowl, however, that intrigues collectors today. The bowl is made of catlinite, a soft, red-colored stone named in honor of the artist and explorer George Catlin. Catlinite is quarried only in a small area of southwest Minnesota.

European Pipes

After their contact with the Native Americans, the European explorers returned to their homes carrying tobacco, as well as Native American clay and stone pipes. These pipes were copied in Europe and then reduced in size to accommodate the rare and enormously expensive commodity of tobacco leaves.

Tobacco and the use of the pipe spread quickly throughout the world. The Portuguese established tobacco plantations in Brazil and Africa in the early 1500s. Dutch traders brought tobacco into Denmark and the Netherlands at around the same time. England seized its first ship load of tobacco from Spain in 1565. Several years later, as legend has it, Sir Walter Raleigh was enjoying a pipe while visiting his plantation in Ireland. One of Raleigh's servants—seeing smoke bellowing from his master's mouth and nose—assumed he was on fire and dashed him with a pail of water to extinguish the blaze. Despite this mishap, Raleigh introduced tobacco and the pipe to Queen Elizabeth in 1586 and pipe smoking quickly became popular throughout England.

Types of Pipes

Clay Pipes

The first pipes produced in Europe date from about 1573 and were hand-crafted from clay. The clay chosen for pipe production was kaolin, a whitish mineral of aluminum silicate mixed with very fine sand. Kaolin proved to be easily molded yet sturdy enough to keep the fire away from the body and clothing.

Manufacturing centers for clay pipes were established quickly in England and Holland. (In England alone, over 3,000 clay pipe makers have been identified.) Other clay-pipe centers developed in Germany, Belgium, Sweden, Spain, France, and Italy. Manufacturers like Gouda of Holland, Gambier of France, and Charles Cropp and William Blake of England were known worldwide by pipe smokers.

Even in the New World, clay pipe makers abounded. Almost every potter on the East Coast and in the Ohio Valley produced clay pipes as a side business. By the mid-nineteenth century, clay pipes were the most common pipe in America. In fact, clay pipes were so plentiful and inexpensive that tobacco companies offered a free pipe with the purchase of some of their tobaccos. The clay pipe industry would flourish for nearly four centuries—for example, in Tennessee, the last clay pipe was manufactured in 1941.

Overall, the clay pipe is the most common pipe ever to have been made, as great quantities were formed in molds and mass produced. Clay pipes were made in a two-piece mold of wood or iron. The moistened clay was pressed into the mold by hand, removed to dry in the open air, and then hardened in the kiln with intense heat. Most clay pipes were very nondescript in design and decoration. The longer-stemmed "churchwarden" pipes ruled the day in early times, while the simple elbow-shaped or "ell"-shaped bowl and shank were most common in the nineteenth century. The French manufacturer Gambier pro-

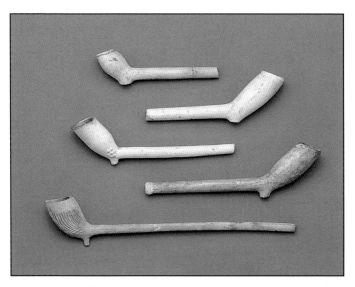

Clay pipes from England, c. 1640 to 1800.

duced the finest and most decorated clay pipes in history. These figural pipes were highly detailed works of art, illustrated with busts of nobles, heroes, and other famous personalities.

Clay pipes were also made and used in other locales. In Africa, for example, tribes extended their tribal art onto pipes. Using local clays tempered with tiny pebbles and shells, the fire-hardened pottery exhibited stylized art of wild animals, geometrical decorations, and human forms. The Middle Eastern peoples were fond of very smooth terra-cotta Chibouks fitted with very long stems. The long stem enabled the smoker to sit on the floor or ground while he enjoyed his smoke. The long stem also piped the smoke over a longer distance, cooling the strong tasting "Turkish" or "Oriental" tobacco.

Wood Pipes

The first wood pipes carved in Europe were copied from both the Native American examples and the first generation of European clay pipes. Some of the earliest recorded wood-pipe production was in Germany in the late seventeenth and early eighteenth centuries. The first commercially produced wood pipes were made by individual artisans or small shops.

Woods selected by carvers included hazel, walnut, wild pear, oak, cherry, apple, boxwood, elm, rosewood, and maple. The artisans carefully chose the species of tree and section of wood specifically needed for each pipe. Trees that grew more slowly and provided more resistance to climate and wind offered denser and harder wood. Trees grown in dry, rocky soils in higher elevations were chosen for their strength and tight grain.

Preparing the wood to be carved was a long and complex process. In early spring before the sap rose, a one-foot band of bark was cut away above the root system and a channel was cut to the core to drain sap. The next winter the tree was cut down. The sections and limbs needed for pipe production were removed and carefully stored for several years for final drying. Finally, the pieces of wood were boiled in water to remove any remaining sap and to improve the hardiness and durability of the wood. With the introduction of steam engines in the mid-eighteenth century, the pipe artisans began steaming their wood as well.

At first, the artisans were self-taught. Soon carving guilds were established, although countries like Germany resisted their formation. Organized guilds and apprenticeships offered education in art and carving and provided necessary experience. The apprentice worked for years perfecting his skill and knowledge before becoming a master carver. Apprentice work included cutting out the general shape of the pipe, drilling holes for the bowl and stem, and sanding and polishing the completed pipe. The master carver would then complete the carving of the pipe, including shaping the ornate details and figures around the bowl and shank. The tools of the trade included axes, handsaws, chisels, knives, drills, vises, rasps, sanding materials, and lacquers.

Early to mid-nineteenth century German wood pipes in the Ulmer and Debrecen styles. The Debrecen is U-shaped with a round bowl and the Ulmer has a flatter profile and is U-shaped.

Carving centers developed throughout Germany, Austria, and Hungary. By the middle of the eighteenth century, artisans in England, France, and the Scandinavian countries were also commercially producing wood pipes.[1] These areas were slower in developing wood carving centers because of the immense popularity of inexpensive clay pipes there. France in particular was very slow in bidding farewell to its love affair with snuff.

One of the more popular wooden-pipe shapes was the Ulm. A flat U-shaped pipe rounded and wide at the mid-section enabling it to be carried in the smoker's pocket, the Ulm originated in Germany. The shank of the Ulm was topped with a short, stubby push stem. Additional features of the Ulm pipe included metal windcaps, many enhanced with ornate cutwork, and metal chains attached to the stem and windcap. Later the stems became longer and were made of turned wood, bone, or antler. Finally, these pipes were finished with careful polishing or were embellished to various degrees with carved scroll and floral designs as well as depictions of events, people, and townscapes from Greek and Roman mythology.

Porcelain Pipes

Another pipe material, popular from the late eighteenth to early twentieth centuries, was porcelain. Porcelain pipes were decorated with hand-painted pastoral scenes, battle depictions, beautiful ladies, and scenes of city life. Later, transfer printing was substituted for hand-painting to save labor and time. Most porcelain pipes have long stems of wood, particularly cherrywood, ebonized bone, and horn. These pipes were made by the thousands, given as gifts to retiring German military personnel, and sold as souvenirs to tourists.

The German regimental porcelain pipes are especially striking. Decorated with multicolored tassels, cords, and military motifs, these pipes commemorated a soldier's service. On the reverse side of the bowl in beautiful scroll writing are the names of his comrades. Most of these regimental pipes have stems averaging two to three feet in length, while an occasional pipe can be found with a stem of six feet or longer. How do you suppose a gentleman lit that pipe?

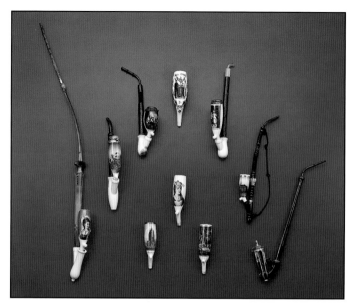

Porcelain pipes were made throughout the nineteenth and early twentieth centuries in Central Europe.

Meerschaum Pipes

The meerschaum pipe marked the pinnacle of the pipe carving art. Featuring exquisite portraits and landscapes, the most beautiful carvings in the history of pipes are found in meerschaums.

Meerschaum, a German word meaning "sea foam," is a very light, clay-like mineral mined mostly in Turkey, near the village of Eskisehir. Meerschaum is also found in many locations in the United States including South Carolina and Pennsylvania, but its quality is not suitable for pipe making. The scientific composition of meerschaum is hydrous magnesium silicate, not fossilized sea creatures as many people believe.

Many legends surround the origins of the meerschaum pipe. One story has it that around 1725 a Hungarian cobbler named Kovács had the good fortune to carve a pipe out of a lightweight whitish "stone." The cobbler, having cut the stone into two sections, proceeded to carve the pieces. Once he completed his work, he noticed one pipe had a waxy finish. Another version of the story has it he accidentally stained the pipe with shoemaker's wax, while yet another version says he may have inadvertently dripped candle wax onto one of the pipes. In any event, the pipe afforded him the finest smoke of his life![2]

The Holy Roman Empire—present day Germany, Austria, and Hungary—was home to the first carving centers for meerschaum. By 1800, the city of Ruhla, for example, had twenty-

The earlier meerschaum pipes tended to be U- or L-shaped with relief carving and flexible push stems. The above examples were carved in Eastern Europe and Austria in the mid-nineteenth century.

seven factories and 150 carvers working meerschaum. Shortly, Vienna, Austria, became the meerschaum carving capital of the world.[3] In the second half of the nineteenth century, meerschaum centers could also be found in Paris, Leipzig, London, Prague, and New York City.

Artisans working individually or in groups developed skills and knowledge that were considered proprietary. Shops developed formulas for pre-coloring meerschaum bowls to reduce the break-in time. The wax-finishing process was another well-guarded secret that would go to the grave with many of its creators. Not even family members were privy to some of the concoctions tradesmen developed to finish their pipes. As a

These three figural meerschaum pipes beautifully illustrate the pinnacle of art found in pipe making and the degree of detail inscribed in meerschaum.

result, today's meerschaum carvers and restorers are often unable to replicate past work. In contrast, designs, themes, patterns, and samples for meerschaum pipes were often available to carvers from pattern books.

Meerschaum pipes went through several transformations from the early 1700s to the early 1900s. Their first phase consisted of large U- or L-shaped pipes or "lap" pipes. Popular until about 1850, these pipes were usually finished in one of three ways: smooth; baroque- ornate carvings including scrolls, festoons, and leaves; and bas-relief carvings of people, battles, and buildings.

The second phase was influenced greatly by the clay pipes produced in France. Beginning in the mid-nineteenth century and continuing until the early twentieth century[4] meerschaums became smaller and the bowl increasingly took the form of a bust of a person or animal. The stems became more lateral as well, and were usually made of amber.

The third phase was greatly influenced by the popularity of the cigar and the cigarette. Known as cheroot holders, these smaller representations, often resembling a tube pipe or an early clay pipe, were topped with three-dimensional carvings of deer, horses, or dogs.

The final phase of meerschaum pipes came about as a means to compete with the growing popularity of the briar pipe. Meerschaum pipes, like briar, became void of decoration and ornate carving. The meerschaum pipe had lost its grandeur and was reduced to a smoking tool once again.

Briar Pipes

The emergence of the briar pipe in the nineteenth century ended the reign of the meerschaum. Pipes changed to reflect the modernization impulse of the twentieth century. Life in the modern age was no longer conducive to fancy carvings on fragile pipes or rooms littered with spittoons, smoking chairs, and pipe cabinets.

Like the meerschaum pipe, the origin of the briar pipe is surrounded in mystery. One story tells of a Frenchman, who visited the birthplace of Napoleon Bonaparte in Ajaccio, Corsica. This gentleman accidentally dropped and broke his cherished meerschaum pipe. Searching the area for a suitable wood to replace his meerschaum, he found a piece of "bruyère," and a local pipe carver fashioned a new pipe for him out of the briarwood. The wood selected by the Frenchman was from the burl—or outgrowth—found at the base of the heath tree. The legend goes on to say that he returned to his home in Saint-Claude, France, a major wood-pipe carving center, with samples of the briarroot. When Saint-Claude began producing briar pipes in the 1850s, the town's quaint carving shops and factories were already trained and equipped. No retooling or major preparation was needed for the new material. Saint-Claude, therefore, became the manufacturing center for briar pipes and would remain its capital until the early twentieth century. Even today, there are still carvers in Saint-Claude producing traditional briar pipes.

Similar in design to the heavily decorated French clay pipe busts, the early briar pipes were smaller than the earlier German and French lap pipes and the porcelain-like wood carvings. The styling of early briar pipes exhibited superb detailing and featured carved busts and faces filled with realistic emotions. Crafted by the French, the more practical and versatile briarwood offered the smoker a lighter-weight pipe with a simple, short, push stem of amber, vulcanite or wood.

The era of briar pipes can be divided into three distinct carving phases, the last two of which overlap. The first phase lasted from the 1850s until the dawn of the twentieth century. Phase

Some of the finest carved figural briar pipes came from Saint Claude, France, near the end of the nineteenth century. In an attempt to capture some of the meerschaum market, the briar carvers excelled in brilliant carving skills.

two includes the time before World War I and continues through today. The final phase began in the 1950s and 1960s when Danish carvers invented the freehand, a style that is still very popular today.

The first era of briar carving was more of a transition from porcelain and turned-wood pipes. These early briar pipes have large bowls, sometimes with reservoirs to capture the tobacco juices, long push stems, and horn, bone, or vulcanite mouthpieces. Another variation of the early briar pipe would last much longer than the large wood ensemble. Copying the figural clay pipes popular at the time, many of the early briar pipes were adorned with carved busts of people and animal heads. By the 1880s, however, the pipe had become very streamlined and in many instances still looked like a clay pipe only with more

During the formal years of briar pipe making, pipes took on many shapes and forms. These late nineteenth century briar pipes from Saint Claude, France, resemble clay pipes of an early period but are adorned with decoration common to the late 1800s.

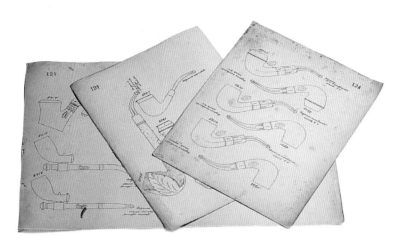

Early line drawings of briar pipes from a 179-page portfolio illustrating approximately 1,100 briar pipe designs from Saint Claude, France, c. 1891.

detailed ornate features. Hand carved briar pipes featured floral decorations of vines, leaves, acorns, and insects on rusticated pipe bowls and stems resembling tree trunks, limbs, and bark. By the time the Art Nouveau movement had fully matured around 1900, Saint-Claude briar pipes were being masterfully crafted with graceful curves, motifs from nature, and life-like busts.

The second period of briar pipe carving began in the early twentieth century when the pipe began to exhibit simpler designs with clean lines. Like other artistic movements of this era, pipe carving had divorced itself from ornamentation and oversized presentation. Classic pipe shapes such as the apple, billiard, Dublin, and poker became the mainstay in pipe design because of their clean lines and simple forms.

Pipe bowls became smaller and stems shorter to allow the modern man behind the desk, on the line in a factory, or behind the wheel of an automobile to continue his business while he

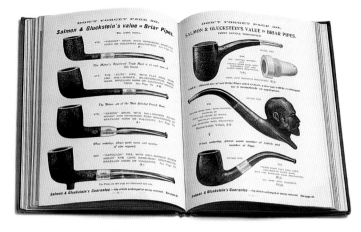

Briar pipes from the catalog, Salmon & Gluckstein's Illustrated Guide for Smokers, *London, January 1899, pp.72-3.*

enjoyed his daily tobacco. In contrast, pipes prior to the twentieth century were viewed both as tools and as artistic endeavors, and were appreciated as such by a more relaxed and slower-paced society. The smooth lines, rounded bowls, and clean finish of the more modern pipe soon spread into Italy, the Netherlands, Denmark, and the United States.

The third period of briar pipe carving came as something of a reaction to the traditional carving standards of the previous fifty years. Freehand challenged the exacting parameters on shapes and dimensions of the earlier periods. Not surprisingly, it occurred in the context of the social revolutions of the 1960s and 1970s. The pipes from this era, which are still very popular, imply freedom, self-expression, and independence. The freehand carver ventured to go where no pipe carver had gone before, following the grain of the burl and allowing it to dictate the design of the pipe. The abstract nature of the freehand pipes appeals to people's individualist attitudes, as no two pipes are the same. Asymmetrical designs, fluid movements of angles and lines, and sporadic dabbling of natural burl surfaces or rustication all add to the appeal of the freehand.

Pipe Production

Throughout the history of pipe making, the pool of people involved in the production of pipes expanded outside the bounds of the carvers. The pipe trade has included skilled and unskilled workers from many occupations. Metalsmiths were employed to cut, hammer, cast, and engrave windcaps, shank rings, inlays, and decorative chains made of silver, gold, brass, iron, and other metal alloys. Artisans of staghorn, ox horn, and bone made stems, inserts, mouthpieces, and other decorative features. Skilled workmen cleaned, colored, cut, and turned horn and bone as the final touches of a pipe. Early meerschaum, wood, and porcelain pipes used ebonized bone (bone that has been colored or dyed black to resemble ebony) as a substitute for the more exotic and rare ebony. The call was also extended to carpenters who crafted beautiful wooden boxes for special presentation cases or companion sets for pipes, extra stems and mouthpieces, and pipe tools. Most meerschaum and briar pipes and meerschaum cheroot holders were sold with small, fitted cases requiring the labor of tradesmen in wood, leather and textiles, as well as metal workers. Jewelers were needed to cut precious and semi-precious stones for mounting on pipes and windcaps. Finally, artisans of amber, vulcanite, and hard rubber were gainfully employed to make stems for pipes.

Other peripheral trades involved in pipe making included glass blowers and potters for the production of glass pipes, clay pipes, and china bowls and reservoirs for porcelain pipes. Glass pipe production was centered in Bristol and Nailsea, England and Venice, Italy in the late eighteenth and early to mid-nineteenth centuries. These magnificent works of art included long swirls of colored

Glass pipes were designed as an advertising medium for the tobacconists and glassmakers shops. The clear-and-white ribbon glass pipes are from Nailsea, England, and the cobalt blue glass pipes are from Bristol, England. c.1780 to 1830.

Snake and puzzle pipes c. 1780, from Staffordshire, England, were smokable but were better known for their whimsical nature.

and clear glass measuring from a few inches to several feet in length. The glass pipes were not designed for smoking, but rather served as trade signs for the glass industry and tobacconists.

During the same time period artisans in the Staffordshire area of England made whimsical pipes of long hollow tubes of clay by fashioning coils, loops and figures of snakes, puzzles and personalities. Unlike the glass pipes, these multicolored whimsies were smokeable. Because of their ornate design and beautiful hand coloring, the "snake and puzzle pipes" were used primarily for special occasions. When not in use, the Staffordshire pipes adorned people's homes as chimney ornaments.

Part II.
The Art of Pipe Making

This briar pipe is made by Joan Saladich of Spain, one of the finer figural pipe carvers in the world.

The Materials

Pipe carving today is practiced around the world. Today's industry is varied, ranging from one-person shops to medium-sized factories. Most craftsmen make pipes as their primary vocation. Some carvers are hobbyists, while others carve to generate supplemental income.

Today, most pipes are made from briar, meerschaum, corncob, and clay. Of the four types, briar is by far the most common pipe material. Corncob pipes or "Missouri Meerschaums" are made in large quantities but are primarily sold as novelty items. The few that are actually smoked are appreciated by their owners as producing a cool, sweet smoke. Clay pipes, in contrast, are purchased more for their history. Some clay pipes are smoked but they emit an earthy flavor. In addition, clay pipes are easily broken and are not practical for travel. Meerschaum pipes are considered the most beautiful by many smokers and provide a cool, neutral, or sweet smoke.

The artisans we will meet in the next section carve pipes of briar and meerschaum. Briar and meerschaum carvers share many similarities including artistic talent, technical skills, carving tools, and low pay. Pipe carvers as a whole tend to be underpaid for their creations and in many instances, even respect or an understanding of their work is dreadfully lacking. To comprehend more fully the tasks of pipe carvers worldwide, let's examine how block meerschaum and briar are transformed from a raw, natural state to a functional and artistic pipe.

Meerschaum

Meerschaum is mined near Eskisehir, Turkey, about 200 miles southeast of Istanbul. Meerschaum is becoming increasingly difficult to obtain, as mining for the "white gold" is hard, back-breaking labor. Meerschaum nodules are brought up in buckets from mine shafts extending up to 400 feet below the surface. Workers remove the clay from the nodules, and then use meat cleavers or hatchets to chop off the harder surface of the nodule which is embedded with sand, dirt, and pebbles. The carvers or apprentices then saw or chop the meerschaum blocks into sections and rough out the basic pipe shapes.

The carver next soaks the meerschaum in water to make it pliable for carving. (Dry meerschaum does not cut cleanly causing small fissures to form on the surface.) The artisan then retrieves the block from the water and carefully trims away with small knives and gouges. As he sinks the blade into the soft meerschaum, he removes long slivers of the white material. The process is somewhat similar to cutting or "carving" a block of cheese. After one or two minutes of carving, he returns the block to the water for another dose. Because meerschaum is so porous, the water is absorbed quickly into the block and it evaporates just as quickly.

After the pipe has been carved and the holes drilled, the apprentice or worker begins to sand the pipe. Using very fine sandpaper and burnishing tools, he will spend one to two hours smoothing out the tiny ridges left by the carver's knife. The pipe

A block of meerschaum.

is then dried for a few hours in an oven. This removes any moisture remaining in the pipe. Finally, the pipe is boiled in a beeswax solution. This last task allows the liquid beeswax to be absorbed into the meerschaum, sealing the pores. The beeswax provides a harder surface for the pipe. It also causes the heated tobacco juices to be absorbed into the pipe resulting in highly desirable golden honey or deep mahogany finishes.

Figural pipes or ones with ornate designs require anywhere from several hours to a full day of carving by the master artisan. On larger meerschaums, the artisan will spend several weeks of carving before completing the pipe. One pipe from the former

Au Pache of Nancy, France carved this massive meerschaum pipe of "Diana", from 1854 until 1869. The pipe with its amber stem weighs 7½ oz. and reflects the style that would dominate the meerschaum market for the remainder of the century.

Museum of Tobacco Art and History in Nashville, Tennessee was carved by Au Pache over the course of fifteen years. From 1854 until 1869, Pache meticulously carved a thirteen-inch, seven-and-one-half ounce masterpiece meerschaum pipe of "Diana."

The meerschaum carver's arsenal of tools is usually handmade and includes small knives with long, slender handles, hand chisels, U- and V-shaped gouges to cut curves and angles in the soft material, and burnishing tools to smooth areas impossible to reach with sanding paper.

Briar

Briarwood comes from the heath tree or bush (Erica arborea). The short, scrubby evergreen is found only in the arid climate of the Mediterranean Sea area. It is harvested in Corsica, Italy, Greece, Spain, Morocco, and Algeria. Pipe makers must estimate their anticipated pipe sales for one, two or three years into the future to insure an ample supply of briar on hand for pipe production.

The wood for making briar pipes comes from the burl or protuberance found at the base of the heath tree. The burl, whose top can be seen protruding above the ground at the base of the tree trunk, has functions similar to a camel's hump; it stores nutrients and water for the dry season. From the burl,

Blocks of briarwood or ébauchon ready for pipe carving.

roots grow in all directions in search of nutrients and water.

Harvesting briar is a chore few people today are willing to undertake. The best briar for pipe making is found in remote rocky soil, usually on mountain tops and steep slopes. Wherever the geographic conditions are at their worst, it seems, one can find high quality burls. In fact the more the heath tree has resisted nature, the denser and stronger the grain of the burl will be. To be suitable for harvesting, the scrubby tree should be at least forty years old and the burl about eight to ten inches in diameter. Once the right specimen is found, the trunk of the tree is cut down with an ax or hand saw. The remaining root system is left intact and will slowly replace the old tree.

The outer part of the burl provides the highest quality briarwood—the straight grain. This area, known as "plateau briar," is cut into large sections. The rough outer surface is left intact. The inside, central area of the burl is cut into smaller sections or "ébauchons." In their appearance, the ébauchons are a lesser-quality wood. Many pipe experts agree, however, that the smoking qualities of the ébauchon are equal to those of the plateau sections.

Plateau grain is usually straight with alternating bands of light and dark wood. The tighter, denser, and straighter the grain, the more desirable it is. When the wood is cut perpendicular to the straight grain, the cut section yields a different grain called "bird's-eye". This beautiful and very tight grain has the appearance of tiny black or dark brown dots surrounded by small swirls of wood.

The ébauchon, being in the center of the burl, has a mixture of grains and can appear to be totally inconsistent as it swirls and changes directions, texture, and color. The ébauchons are cut into small L-shaped sections for the fraising machines. Because the ébauchon grain is inconsistent and burls yield more ébauchons than plateau blocks, ébauchons are much less expensive and are used to make the vast majority of briar pipes in the world today.

Most pipe makers boil the briar in water, oil, or some other guarded brew for ten to twenty-four hours to remove the sap. In the nineteenth century, most manufacturers steamed briar to force the sap out. Removing the sap is necessary because the sap holds minerals and acidic tannins that cause the pipe to smoke bitter or have a foul taste.

The cut blocks—whether plateau or ébauchon—are then dried. Some pipe manufacturers carefully air dry their blocks for one or two years. A small minority of carvers dry their briar for three to five years or longer. The storage areas are usually bins in the factory or outdoors under tarps. At first, water is routinely put on the briar to prevent it from drying too quickly and causing severe cracks in the wood. Currently, there is a great debate about the proper age of cured briar. Some industry heads believe that once the sap has been removed and the briar dried for several months, it is ready to be produced into a top-quality pipe. Other pipe carvers say the briar, like wine, improves with long-term aging.

Much of the disagreement has to do with the economics of

pipe production. Pipe manufacturers invest in briar blocks for several months to several years before a return is realized. In general, the longer the briarwood is cured or dried, the sweeter the smoke of the pipe.

Carving

The final phase of pipe making, the carving process, demands the most planning, skill, and talent. This phase is also the most open for interpretation and imagination. No two artisans carve alike, nor do they approach the briar in the exact same manner. There are three major production philosophies for carving briar pipes, although they overlap and can easily blend with one another. This is what makes the briar pipe production so interesting. Each artisan has his or her own way of crafting pipes. Some techniques and approaches are better than others. Some modes are very expensive and other systems may not bring out the finest in the briar. Only the smoker can decide on the techniques, styles, quality, and price that best fits his eye, desire, and wallet.

There are no two identical pipes in the world. Even if the design, shape, size, and raw materials are all the same, and the pipe is carved by the same hand or machine, there will still be differences. The subtleties may not be noticeable to most, but both the smoke and the grain will be different to the most discriminating connoisseur.

The art of pipe carving involves many steps, as well as precision. The final steps include sanding, staining, polishing, and gently bending the stem with heat.

From the beginning of briar pipe history, pipes were turned mechanically on lathes powered by steam engines. By 1900, when Saint-Claude was at its peak, annual production reached fifty million pipes of which the large majority were shaped by machines. Most pipes made today are turned on fraising machines. Fraising machines automatically bore and shape the pipe. One machine will bore the bowl chamber, mortise hole, and air hole. Another machine will shape the outside of the bowl. Then yet another machine will turn the shank while a final machine trims the foot of the bowl. In all, making a briar pipe requires 80 to 120 steps. This is the production norm of the industry and is accepted by the majority of the pipe smoking population.

This is not to say the entire pipe is made by machine. In fact, every pipe made today is handmade. Even though a pipe may be completely turned by machines, people still have to grade the blocks, place them accurately in the machines' clamps, sand away the uncut edges, smooth the pipe with sanders, and complete the process with stain, wax, and buffing. Most would agree, though, that pipes made with a series of fraising machines are correctly classified as machine made.

A machine-made pipe does not necessarily represent a lesser quality pipe, however. In fact, one of the benefits of fraising machines is that exact dimensions of a classic style bowl are consistent. Using a template, the machine fashions a pipe shape that may otherwise be very difficult if not impossible to carve by hand. One difference between a fine quality machine-made pipe and a lesser quality machine-made piece is the ébauchon block. The finer pipes are made from blocks specifically chosen for that pipe, highlighting either the superb grain or a totally flawless piece of briar. Some of the finer pipes made in the world today are shaped by a machine.

Along the same lines, even the very best carvers in the world today use lathes, saws, and sanding machines to cut, bore, trim, and polish their "handmade" masterpieces. The amount of machinery used in producing handmade pipes varies from carver to carver and manufacturer to manufacturer. Some carvers will manually cut, trim, and sand briar blocks on lathes and grinders. Other carvers will use some fraising machines to do a single step or a combination of steps—including boring holes, shaping the bowl, shank or heel—but they will finish the pipe by hand. Then there are carvers who carve and finish the whole pipe by hand with knives, sandpaper, and steel wool.

A brilliantly and masterfully carved pipe by Joan Saladich y Garriga is ready for its trial smoke.

Selecting a Pipe

So what makes a pipe outstanding? Before you review examples of some of the finest pipes in the world and the artisans who have put their hearts and souls into crafting a pipe you can enjoy for a lifetime, let us discuss how to select a pipe. Selecting a pipe is similar to buying a car. There are many cars on the market from the small economy size to a high performance sports car.

The price of pipes in most cases is dictated by whether the pipe was crafted by hand, automatic machines, or a combination of both. Some of the carvers featured in the following pages spend more than ten hours on a single pipe. How does one charge to compensate for the time? James Margroum, known as "Mr. Groum" in the pipe industry, jokingly comments that once the pipe is completed and his time counted, he has earned about seven cents an hour.

Materials

Time is not the only criteria used in determining price by the carver or company. The quality of the briar or meerschaum, added features such as gold bands and unusual or exotic inserts all influence the price of a pipe. In briar, the larger-sized blocks with straighter and tighter grain command a more expensive price. Quality straight-grain plateau briar blocks range in price from ten to thirty dollars a piece. Some artisans, in pursuit of the very finest straight-grain plateau briar blocks, known as "premium, premium," are paying as much as $100 a block sight unseen.

One must keep in mind that the market drives the prices as well. The ébauchon blocks provide a quality smoke, but the more serious smokers and collectors prefer the highly visible straight grain bowls. Many pipe makers will argue, however, that the ébauchon bowls perform equally as well as the straight grain bowls in fine-smoking pipes.

Price fluctuations are also the result of other aspects of the production process. For example, most carvers and factories use stock or pre-formed stems requiring little or no additional work and time by the tradesman. Only a few carvers in the industry actually make or cast their stems and then handcut the shape to conform with the pipe's design. In fact, some of the finest artisans spend as much time on stem design and construction as they do on the pipe bowl.

Finish

Another point to keep in mind is how the pipe is finished. Many machine-grade pipes and some hand-carved pipes may not have high quality stains and waxes. Each shop has its own formulas and techniques. Secondly, many machine-made pipes and hand-carved pipes have rough edges or exposed putty marks covering flaws or sandpits. Most plateau and ébauchon briar blocks have minor flaws and cavities, and almost all pipes have some small blemishes. In many instances, the craftsman can remove the blemish and not harm the quality of the pipe. In other cases, the flaws cannot be removed without altering the design or compromising the pipe. These situations require the area to be filled or camouflaged, and the price of the pipe should reflect this alteration. On less expensive pipes, the putty fill can be very noticeable.

Most sandblast and rusticated pipes are finished with a rough texture to hide sandpits or other flaws. A sandblasted or rusticated surface does not mean the pipe is inferior. Another point to consider is that while sandblasted and rusticated pipes are less expensive, the quality of smoke is not impaired. If anything, it might be enhanced because the textured surface allows the pipe to smoke cooler. Remember as well that the craftsman has actually given these pipes much more of his time and energy than a standard smooth pipe. In that sense, it should be worth more than a smooth one. In general, textured pipes are a great way to own a master carver's work without paying premium dollars.

Other blemishes and flaws can include tool and sanding marks left unfinished, holes drilled off-center, improperly fitted stems, air passages that do not draw freely, design elements or features that are not in proportion with the pipe, and poor pipe designs. Each blemish, flaw, or mistake can and should affect the final price of the pipe.

Machine-made versus Hand-made

Most pipes purchased today average forty to fifty dollars. Pipes in this price range are machine-made from ébauchon briar. Yet the smoke they provide is quite good. As Frank Blews of Lane Tobacco points out, an occasional machine-made pipe will smoke every bit as good as his expensive Dunhills. Inexpensive pipes are inexpensive because the work is automated and the materials are standardized. You may find that the stems do not always fit tightly or that the pipe has very noticeable putty fills.

Most of the mid-range pipes, costing between one hundred and four hundred dollars, are hand-crafted. A minority are shaped on fraising machines. All use high-grade briar, have a beautiful appearance, smoke well, and may or may not have noticeable blemishes. In this price range, you will find the finest names in pipe making, and any pipe in this range is a quality pipe. Annual sales of pipes in this price range are about 50,000 pipes.

Finally, there are the premium pipes, the Chonowitsch, S. Bang, Ivarsson, Cooke, and Nordh pipes and the high-end pipes made by Castello, Dunhill, and Savinelli. These are pipes that achieve the highest levels of design, appearance, and performance. These pieces of art can cost ten thousand dollars. Annual sales in the United States of high-end pipes might number 1,000 bowls.

Features

The most important feature of a good smoking pipe is a free air passage from the bowl to your mouth. As you examine through the endless examples of pipes at your favorite shop, check the airflow by inhaling the pipe as you would if you were smoking. The passage should be clear and the airflow unrestricted. A whistling sound may indicate an obstruction within the air passage or a joint that is constricting the airflow. Any restriction of the airflow will make it difficult to keep the pipe lit. In addition, because you have to puff harder to keep the tobacco alight, heat will build up rapidly in the pipe and result in an uneven burn of the tobacco.

Other major considerations in purchasing a pipe are balance, weight, and bowl size. How a pipe balances in the mouth is important. Avoid pipes that are too heavy or out of balance. A poorly balanced pipe or a heavy bowl can lead to sore jaws and teeth. A new pipe in a shop may not seem heavy or unbalanced at first, but handling it for a short period of time before purchasing should give you an indication of how the pipe will function in the mouth.

Bowl size is equally important. People have different requirements for bowl sizes and shapes. Many people prefer larger bowls because they smoke cooler. Others favor small bowls because the pipe fits into their pockets or purses. Larger bowls weigh more and can unbalance the pipe as well as weigh on a jaw. Larger bowls are often enjoyed by taller or larger people while Europeans, ladies, and men of smaller stature tend to cater to smaller bowls. Conversely, conventional wisdom has it that round-faced smokers prefer straight stem pipes and long, narrow-faced folks tend to smoke full-bents. The list goes on and on of matching personalities, sizes and shapes of pipes to people. In short, choose a pipe that appeals to you, feels right in the hand, complements your facial features, and looks handsome to smoke.

Smoking a Pipe

The smoking quality of the pipe depends on several related elements. First, the briar must be well aged with most, if not all, of the sap removed. Second, the air hole must precisely meet the base of the tobacco chamber. Third, the air hole and stem must have a wide, clear passage. Keep in mind that one block of briar will smoke slightly differently than another block. The grain in the briar controls the breathing qualities of the wood, giving the pipe its characteristic smoke.

Smoking a new pipe can be a wonderful experience, a mediocre experience, or even a horrible experience. How a virgin pipe smokes on the first bowl depends greatly on how the wood was cured and somewhat on the pipe's manufacture. Some of the finest pipe artisans add an oil curing process to ensure a proper tasting pipe. Others may boil the wood or wipe it with oil. In any event, the more thorough the curing of the wood, the better it will smoke from the first bowl. So how can you tell if the wood has been well prepared? The sections that follow on the carvers indicate, in most instances, how they prepare their briar. You can also inquire from the store attendant about the maker and the manner in which the pipe was crafted.

Breaking-in a New Pipe

Every maker and smoker has a different opinion about breaking-in pipes. Some people suggest wiping the inside of the bowl with honey before filling it with tobacco, while others prefer to use a dab of water then let it dry, and yet others prefer filling the pipe with a quarter or half of a bowl full of tobacco. The key is to not smoke the first several bowls too hot. Keep the pipe well lit, but smoke slowly until all of the tobacco has been consumed. Otherwise, the extreme heat could crack a new bowl.[5] Inside bowl treatment will vary from pipe maker to pipe maker. Dr. Grabow pipes, for example, are presmoked and have a thin layer of carbon buildup ready for the new owner. Other carvers such as J.T. Cooke or Jess Chonowitsch have formulated a special coating for the bowl's interior. Some pipe makers use a carbon coating to facilitate the breaking-in process. Many carvers, however, leave the interior in a natural or stained state so the pipe smoker can appreciate and admire the beautiful grain inside and out.

Frank Blews of Lane Tobacco has devised a useful system for loading and smoking a pipe. According to Blews, three ingredients are necessary to enjoy a pipe and tobacco: good quality pipe, good tobacco, and good smoking techniques. Too often, new smokers do not have someone who can patiently show them how to load a pipe and smoke it properly.

Of course, it is also important to purchase a quality pipe.

Many new pipe smokers understandably begin with an inexpensive pipe to save money on their initial investment. Starting with a quality pipe will ensure a smooth smoke that will reward you for your wise purchasing decision.

Choosing a blend of tobacco that is compatible to your palate is important. New pipe smokers tend to purchase a highly aromatic tobacco filled with flavorings. This most often leads to tongue bite. In addition, the heavy moisture in these types of tobacco often causes the pipe to go out. There are many pipe tobaccos available and your tobacconist should be able to guide you in the right direction. Check for an aromatic mixture with a fine cut and not too much moisture.

Many tobaccos are difficult to load into a pipe if the new smoker is unaware of proper loading procedures. You want a consistent pack from top to bottom and this is accomplished by packing in three levels. Begin loading a pipe by lightly packing a pinch of tobacco at the bottom of the bowl. Take another pinch of tobacco and pack it a little firmer in the mid-section of the bowl. The final portion should be packed firmly on top. If packed too lightly, the pipe will smoke too hot. If packed too densely, the pipe will be difficult to keep lit. Either way, the result will be a poor smoke and tongue bite. As you pack each layer, use a pipe tamper to lightly tamp or press the sides of the tobacco forming a dome in the center.

Light the tobacco evenly across the top. Unlike a cigar or a cigarette, you may need to re-light the tobacco several times until the tobacco burns evenly. After lighting, draw on the pipe for a few seconds, tamp down lightly, and then re-light. Continue until you build a white ash across the top of the tobacco. Now lightly tamp or press along the edges of the tobacco by angling the tamper to maintain the dome. Make sure the top of the tobacco is fully lit.

By tamping the sides as you smoke, you force the fire to burn evenly across the top and down. If you tamp often, your pipe will stay lit longer and your smoke will be cooler. If the fire goes out or you set the pipe down for several minutes, re-light and continue your smoke. Remember to lightly draw or puff on the pipe. If you inhale too fast or too hard you will cause the fire to burn too hot and damage the pipe. Pipe smoking requires patience and it is a great stress reliever. It will slow you down and perhaps make you more methodical and analytical.

After you have finished smoking your pipe, clean out the ashes but do not disturb the carbon or cake buildup around the inside of the bowl. The carbon lining actually protects the wood from the fire and provides a cooler smoke. Maintain a maximum cake thickness of about an eighth of an inch. You may prefer to purchase a reaming tool to assist in removing excess cake. A knife is not an appropriate tool because the blade leaves an uneven cake wall and you risk gouging the bottom of the bowl.

Ashes can be removed with a pipe tool. A cotton and wire pipe cleaner bent in half and twisted performs nicely as a swab. Another pipe cleaner can be used to swab the inside of the stem and shank. Remove the stem and clean it; then clean the shank and air passage to the bowl. It is important that you remove all of the moisture collected here, but do not reassemble the pipe until the stem and bowl have cooled down. Now put the pipe away for another day. A pipe needs a day of rest before you smoke it again. In fact, pipe smokers own many pipes so they smoke a different one each day. If you smoke your pipe day after day, the bowl may crack or burn through. Finally, store your briar pipe with the bowl down and the stem up, this allows for any moisture to settle in the bowl.

Remember your pipe is both a tool and a work of art. Do not beat it against hard surfaces to dump ashes. This is the quickest way of damaging or destroying your investment. The shank area, stem, and top of the bowl are fragile elements that require proper care.

So sit back, relax and enjoy your pipe. You will thoroughly enjoy the hunt for other pipes to add to your collection. Happy Smoking!

Roy Roger Webb's finely detailed carving is a work of art.
The briar wood is one of the harder woods and requires great skill and steady hand to create minute detail in his spirit faces.

Introduction

Professional pipe artisans can be found in almost every country in the world. Perhaps 300 to 400 craftsmen are carving pipes today. When a master carver retires or death calls him or her from the workbench, a new talent is ready to join the ranks.

As you examine the different carvers around the world and their unique interpretations, you'll notice similarities and schools of art. The oldest briar school, for example, is centered in Saint-Claude, France. Italy has two distinct schools of carving, while the other Scandinavian countries have a look all their own. Then, there are the United States and Japan. The variety of styles and techniques in the United States reflect America as the proverbial melting pot. In Japan, styling can be very Japanese or exhibit a combination of Japanese and Danish features.

Modern briar and meerschaum pipe styles can be divided into two broad categories: classic and freehand. The classic style has its origins in the early French and English pipes. The proportions of the bowl and stem: length, height, and diameter are standardized in these pipes. Many of the names for classic pipe shapes and styles were chosen many generations ago and reflect the shape or appearance of the pipe. For example, the profile of an "apple shape" pipe very closely resembles the outline of an apple. Other shape names seem to have very little to do with the actual shape of the pipe. Classical pipe shapes include the Canadian, billiard, poker, bent, and Dublin.

Freehand was a term coined for the "wild" or "organic" designs of Danish pipes that first appeared in the late 1950s. Freehand implies that the pipe is carved at the whim of the carver. A truer definition is that the pipe design is governed by the grain of the wood. No two freehand pipes are carved alike, nor do they conform to any one style.

The strenghth and beauty of the briar grain is brought out handsomely in Higuchi's pipes. His styling and depiction of the briar are Danish carryovers.

Canada

Canadian carving techniques and styles are very individualized. Some styles are very traditional and spring from European roots. The Canadian carver Julius Vesz, for example, produces pipes in the Old World tradition. Other carving styles mirror the Danish freehands.

J. Calich

Based in Mississagua, Ontario, John Calich is one of the finest carvers in Canada. He has a talent for crafting pleasing designs and bringing out the beauty of the briar.

Calich has carved pipes for more than forty years; for the last twenty-five it has been his primary vocation.

Calich is a self-taught pipe maker. He has studied at length the work of European pipe masters and American carvers. He strives to make each pipe better than the pipe before, and he runs his business on the premise that "the material and workmanship should speak for itself." To maintain quality craftsmanship and superb design, John uses only the best materials available. He is very critical with his pipe designs and is exceedingly careful with construction techniques and details.

*Calich makes a very fine classic-shaped pipe.
His freehands are very conservative with gentle curves.*

J. Calich pipes begin with plateau and ébauchon briar from Greece, Corsica, and Italy. With straight-grain plateau briar, Calich studies the wood and designs a pipe shape that complements the block and grain pattern.

The tools Calich uses to carve his pipes include the lathe, drills, chisels, rasps, files, abrasive cloth, buffing wheels, and compounds. Some of his most frequently used tools, like the specially contoured sanding wheels, he makes himself. These unique tools allow him to cut, groove, and contour the edges and surfaces of the briar to give it the "J. Calich look." His stems are made from vulcanite and acrylic and are decorated occasionally with inserts of bamboo, exotic woods, and silver and gold bands.

J. Calich pipes are unique, stylish, and have good lines, shapes, and design features. His use of subtle contours and rounded ridges enhance the feel and appearance of his design. He carves in the classic style, but also produces freehands with conservative overtones. Calich's attention to detail is evident in the bowl shapes, air hole and mortise drilling, and the manner in which he brings out the best in the grain. His ability to enhance the flame in the grain is exceptional, as is his eye for selecting the finest briar.

He prefers a larger bowl, but its weight is not that noticeable. His choice of stain colors is very appealing and his finish is very smooth to the touch. Each pipe is stamped "CALICH" and a single, tiny silver dot is applied to the top of the stem. Retail prices range from $50 to $1,000 and annual production is approximately 500 pipes. His high-end, straight-grained pipes number only about 75 to 100 bowls annually.

J. Calich pipe.

Pipes made by Julius Vesz are conservative freehands with classic styling handsomely mixed with Old World traditions. Many of his pipes are oversized or very small like this raindrop.

Julius Vesz

Julius Vesz of Toronto is a master carver who draws on Old World craftsmanship and traditions to make his beautifully carved pipes. Born in Hungary in 1933, he was "intrigued and fascinated" with carving from an early age, as he watched his grandfather carve meerschaum pipes for hours.

Vesz immigrated to Canada in the late 1950s. Trained as a mechanical draftsman, he sought employment but found only part-time work. Because Toronto had so many pipe smokers, he built a pipe restoration business. Vesz was named the official Canadian pipe repairer for Dunhill, and later he acquired the Dunhill title for all of North America.

Vesz is one of only a few craftsmen who work totally in dead-root briar. Dead-root briar is no longer available, but in the past it was the wood that had been killed and dried by nature. Overall, it is a superior briar that offers a sweeter and drier smoke.

Vesz begins with a block of briar that has the appropriate size and grain for the pipe shape he has in mind. He soaks the wood in water to study the grain and then marks the outline of the pipe with a pencil. "The grain always has to complement the shape of the pipe," he notes. "That is one of the priorities I have in my pipe making." He next roughs out the shape with a bandsaw and then shapes and finishes the pipe with sanding wheels. On special pipes, he hand cuts the stems from resin rods. The grain in the resin is beautiful like the wood.[6]

Julius Vesz sculpts an outstanding pipe and he offers a marvelous diversity of shapes. Most of his pipes are traditional but he is also fond of giant, conservative freehands. Many of his pipes exhibit motifs and designs from Eastern Europe. He uses a variety of insert materials for the stems including meerschaum, bamboo, and amber. Not only is Julius Vesz a master carver, he is a master metalsmith. He handcrafts his own metalwork including hammered designs on gold and silver bands. One of his more popular pipes is the small raindrop, a full-bent pipe. He also produces pipes with a beautiful rusticated finish. Julius Vesz makes about 250 pipes a year and retail prices range from $200 to $8,000.

Corsica

Although the island of Corsica has produced wonderful briar for almost one hundred and fifty years, L.J. Georges, a descendant of a briar exporting family, is the only pipe carver in Corsica.

L.J. Georges

Victor Georges began Riviera Briar Pipe Company in 1920 to export briar blocks from Corsica. The company sold about one million blocks each year until it was sold in 1975. L.J. Georges, Victor's grandson, decided he wanted to continue working with briar after the family company was sold. He traveled to Denmark to study pipe making with Sven Knudsen for three years before returning home. L.J.—Lucien Jassois—carved pipes for several years before he felt his work was worthy to sell.

Today, Georges is the only briar pipe carver in Corsica. His location enables him to be one of the only carvers in the world to harvest his own briar. He searches the mountainsides for the briar burls. After returning with the burls, Georges cuts the wood into blocks and boils it for eleven or twelve hours, and then air-dries the wood for three years.

Georges is one of the few carvers experimenting with exotic woods. The unusual woods lend a handsome appearance and the briar insert provides a cool, clean smoke.

Georges' pipes are classical in nature, but have a Danish influence and in some instances, Danish shapes. He studies the grain in the block and then determines the pipe shape. He cuts away the outer portions of the wood and uses a sander to roughly shape the pipe. Only on some pipes will he draw the shape of the pipe on the block; more often he cuts and shapes at will. He uses a lathe to drill the tobacco chamber, shank, and air hole before finishing the pipe with both machine sanders and hand sanding. Georges handcrafts his stems from ebonite and boxwood and he uses horn rings as inserts in some of the pipes. Although Georges mixes his own stains, he leaves some pipes unstained for a natural finish.

L.J. also began making exotic pipes in 1997. These pipes have a briar bowl insert and a body made from exotic woods from Central America and Africa, including olivewood, ebony, rosewood, violet wood, and cocobola wood. L.J. Georges produces a quality pipe with a superb straight grain. His work is very visual and offers good balance and smoking quality. His sandblast pipes are finished in black and burgundy.

Until 1998, L.J. Georges pipes were mostly sold in Europe. In 1999, the L.J. Georges pipes entered the American market. Because of the complexity of combining briar with exotic wood, he can only fashion one to two pipes per day, and averages about 500 pipes annually. Georges' briar pipes begin at $200 for rusticates and top at $1500 for the best straight grains. Retail prices for exotic wood pipes range from $400 to $575.

Georges also produces beautiful classic briars highlighting the exquisite grain and superb finish.

Denmark

The countries of Scandinavia have produced some of the finest pipes in the world. Danish styling in particular has influenced pipe manufacturing around the globe. Known as the "Danish Revolution," the freehand movement began in the 1950s and 1960s when Sixten Ivarsson developed a new "look" for pipes as he explored how briar grain and texture can become the focus of the pipe. Ivarsson based his pipe design on the total shape of the pipe, how the bowl, shank, and mouthpiece all come together in unity. In contrast, the traditional style of pipe making treated each part of the pipe as a separate entity.

So, what influenced Ivarsson to pioneer the freehand design? Niels Larsen points to "the furniture designer and architect Arne Jacobsen—the idea of shaping a piece of furniture according to the flow and structure of the materials. When you bring that concept to a pipe, it means shaping the pipe according to the grain of the wood, rather than forcing the shape into a predetermined form and working against the briar."[7] As a result, Danish carvers including S. Bang, Lars Ivarsson, and Jess Chonowitsch (as well as the Swedish carver Bo Nordh) have created a style of pipe that has influenced pipe styles and pipe makers worldwide. Danish carvers carve slowly and deliberately, and they only finish a pipe when it is perfect! Shortcuts in quality or design are not acceptable to them. Their styling is not the wild, flowing "organic" pipes of the 1970s, but conservative designs that emphasizes the texture and beauty of the grain.

Bang's Pibemageri

In 1984, Sven Bang retired as a tobacconist. Although he never carved pipes during his career, he invited Per Hansen to open a workshop in Copenhagen on January 1, 1970. So began the tradition of S. Bang pipes. In 1971, Ulf Noltensmeir joined Bang and Hansen. When Sven Bang retired, Hansen and Noltensmeir took over the firm Bang's Pibemageri and the pipe brand S. Bang.

Hansen and Noltensmeir use only the finest plateau briar blocks available from Corsica. They pay a premium for the blocks and have built their reputation on the extraordinary presentation of the briar grain in their pipes. The "Bang boys" create designs for their pipes by trying "to find the pipe in the briar." Their arsenal of tools includes the "lathe, homemade tools, sanding discs, files, knives, and a lot of sandpaper." Striving for perfection in each pipe, they are very deliberate and precise with their actions. Hansen and Noltensmeir's philosophy of pipe making is to "create as beautiful pipes as possible without compromising with our perception of perfect smoking." They hand craft stems from ebonite and German hard rubber rods. Many of their pipes are decorated with silver, gold, ivory, and a variety of horn and other woods. Their pipes have some of the finest stain and finishes of any pipe in the world.

S. Bang pipes are both very Danish and conservative in design. Their shapes are bold in their simplicity, and both the straight and the birdseye grain are so superbly presented one cannot but want to own, treasure, and smoke an S. Bang. These are works of art that function like a high-performance automobile. S. Bang makes a total of about 500 pipes each year. The pipes are stamped "S. Bang" and the initials of Hansen or Noltensmeir. Retail prices start at $650 and top $3,200.

S. Bang pipes have a distinctive Danish appearance and their presentation of the briar is among the finest in the world. They are known for their beautiful, flowing designs with graceful lines, and are also some of the finest smoking pipes.

Jess Chonowitsch

Chonowitsch learned the art of pipe carving from acknowledged Danish masters Paul Ramussen and Sixten Ivarsson, and began a pipe-making business with his father in a rural village outside of Copenhagen. In 1972, he established his own shop in a small building behind his house in Bråby, Denmark.

Chonowitsch makes both classical and freehand pipes. Using specially designed machines his father owned, he turns the bowls to give them perfect classical shapes. The rest of the pipe he crafts by hand. When he carves a freehand pipe, his approach is very different from other carvers. He creates the design in his mind and then searches for the block of briar that will provide him with the pipe he has imagined. Chonowitsch knows the characteristics of each block of briar he owns. Some of the Corsican and Grecian briar is stored in his shop for ten years, but he knows the location and characteristics of each block and what kind of a pipe it is capable of producing. "Whenever I pick up a new block of briar, I can find a whole world in it. I have to think about the shape, look at the briar and then start. It's never the same," he says.[8] This is why Chonowitsch carves a particular style and works it until he reaches perfection. Then he develops another model and pursues the search.

Chonowitsch buys only the very best plateau briar from Corsica and Greece. When the bags of briar arrive at his Denmark shop, he rejects most of the blocks. Once he has graded the acceptable blocks, he begins the slow process of drying the briar. He will dry the wood for at least one year, but most of his briar has been stored for three to four years before it is selected for carving.[9]

Finding the perfect pipe in a block of wood is a rare circumstance. Chonowitsch begins with the design drawn on paper. He sketches the outline of the pipe onto the chosen block and roughs out the shape on a bandsaw. Additional shaping and smoothing is done with electric sanders. He uses a machine his father had custom made to bore holes for the tobacco chamber, mortise, and air hole. Precision at this stage is what makes Chonowitsch's pipes masterpieces. Chonowitsch finishes the wood with hours of hand-sanding to produce a smooth, high gloss finish.

The final step is making a stem for the pipe. Chonowitsch handcrafts stems from German vulcanite. He drills the air hole very carefully to ensure a perfect alignment and a good draw. The shank and stem are sanded together for a precise fit and the mouthpiece is hand-filed to his specifications. To complete the pipe, he stains the wood and polishes the finish on a buffer wheel.[10]

Chonowitsch makes a remarkable sandblast pipe that beautifully enhances the grain.

Chonowitsch combines aesthetics with perfect function to produce outstanding pipes. Of the 200 to 250 pipes he carves each year, only one or two will be true masterpieces. These very special edition pipes are marked with a bird stamp, in honor of the pigeons living in his yard.[11] All of his pipes are stamped "Chonowitsch Denmark" encircling "Jess." Jess Chonowitsch's pipes retail $700 to $5,000.

Lars Ivarsson

Based in Gelstrup, Lars Ivarsson is one of the giants of pipe carving. The son of the great Sixten Ivarsson—widely regarded as the finest pipe carver of the twentieth century—Lars Ivarsson began carving very early in life. He was nurtured at a young age in the pipe business by his father, and by age thirteen or fourteen he was cleaning pipes and doing repair work.

Sixten Ivarsson began carving pipes at the end of World War II. His designs were radical and exciting, and, as a result, Danish freehands quickly became the new fashion in pipedom. During Sixten's tenure as an outstanding artisan, he trained two apprentices: his son, Lars, and Jess Chonowitsch. He remained a pipe carver until just a few years ago when, at eighty-four, he was forced into retirement by health problems.[12]

Living on a small farm close to nature has greatly influenced Lars Ivarsson's philosophy of pipe design and creation. Each pipe he produces has some relationship to nature. For example,

Ivarsson's pipes are well-balanced and very comfortable to hold. His designs have influenced pipe makers throughout the world.

he studies the shapes and forms of small creatures like the snail, as well as the ebb and flow of the ocean.[13]

When Ivarsson carves a pipe, he already has the design in his mind for the block of briar. As he turns and twists the block in the sander, his subconscious mind controls the movements of his hands. The briar rarely cooperates with his vision, however. Each and every block has hidden traps of sand particles, grain flaws, and cracks.

Lars Ivarsson makes freehand pipes in the Danish style. Like Chonowitsch and Nordh, he is a leader of a movement. Ivarsson places much emphasis on the design and airflow dynamics of his pipes, and strives for perfection with each and every pipe. To reach this level entails precise craftsmanship that few carvers in the world can achieve. In Ivarsson's attempt to produce only the best, he will craft less than sixty pipes per year. If one of his creations has even the smallest flaw as he approaches the finishing stages, it becomes firewood! When he crafts a pipe that, in his opinion, is truly perfect, he marks it with a special stamp of a blowfish, his symbol for a grand masterpiece. A perfect masterpiece from Lars Ivarsson only happens once every year or two. Owning and smoking a blowfish-stamped Lars Ivarsson pipe is the ultimate pipe experience. Retail prices range from $900 to $5,000.

Jørn

Jørn Larsen of Copenhagen has been carving pipes professionally for about sixteen years. Trained in the Danish tradition under Jess Chonowitsch, Larsen has been carving his Jørn pipes for about six years.

Using only the best plateau briar from Corsica, Larsen begins with a design or "model", which he has prepared and draws the design on the briar block using a template. He

Jørn Larsen is a mechanical engineer who has applied his training into pipe making. He makes an incredible smoking pipe with strong Danish lines.

removes the outer edges with a bandsaw and then bores the tobacco chamber on a lathe. He refines the shape with a sanding disc and then using the lathe, cuts the mortise hole and air passage. Larsen makes his pipe stems from vulcanite rods by turning them on a lathe. The stem shape is refined with a file and then sanded by hand.

Larsen handcrafts a very attractive pipe, and his technical abilities are among the stronger points of his work. The mechanics of a Jørn pipe are superb, the air holes and tobacco chamber are perfectly drilled. This is very important to him because, like most Danish carvers, he intends to make a perfect pipe.

Most Jørn pipes are classic in shape and exhibit a beautiful grain pattern. Some pipe shapes have English overtones but, overall, the Danish styling is very evident. Larsen is very careful in presenting the finer qualities of the briar; the grain is pronounced and vibrant. He makes about 300 pipes annually and retail prices range from $340 to $645.

Anne Julie

Based in Laeso, Anne Julie is both a pipe maker and a painter. Her work is noted for its bold form, strong lines, and robust curves. Today, she sells her highly sought after pipes in Denmark, Japan, Germany, and the United States.

Anne Julie happened into pipe making through her husband, the renowned Danish pipe maker Poul Rasmussen. She had been married for only a short period of time when Rasmussen died suddenly at the age of forty-six from complications of a heart transplant. Only twenty-seven and with a two-year-old son, Julie was forced to make some hard decisions. At first, she planned to sell Rasmussen's carving machines and pipe inventory. While visiting the shop in Østerboro, however, she ran into one of her husband's former apprentices, Hans Nielson, known as "Former." Together, they looked through the shop and found some briar blocks. With Nielson's help Anne Julie began carving pipes. She notes that until then she never handled a pipe because her husband believed pipe manufacture was only men's work. Gradually, however, she "grasped the principle of surface area, the nature of wood, the individual log's innate shape, draughts of the year's seasons, shades, functionalism, etc."

The Anne Julie pipes, marked by a single red and white dot, are made by hand. She uses a large grinding wheel to shape the

*Julie makes a variety of pipe styles.
This pipe is one of her more conservative designs.*

pipes and stems. She finishes most of her pipes in a ruby red and black stain and stamps them "Anne Julie."

Anne Julie pipes include both traditional shapes and expressive freehands. The traditional pipes are stunning, beautifully crafted, and large. The freehands exhibit bold forms, strong lines, and graceful curves. These pipes reflect Julie's philosophy of living life to the fullest. Some of the pipes have a feminine feel while others reflect the rugged, austere, and wind-swept climate of Northern Denmark.

One of the most intriguing aspects of Anne Julie pipes is their metalwork. Julie is fond of using metal windcaps or plates of metal and bands on the pipes. Klans Bjerring Andersen does her silver work. She also handworks the stems with contours, beads, and rings to accent the unique designs of her pipes. Anne Julie pipes are very much worth the effort to locate for firsthand inspection and purchase. She produces a limited number of pipes annually. Prices range from $250 to $3000.

W.Ø. Larsen

One of the oldest names in tobacco is W.Ø. Larsen. Established in Copenhagen in 1864 by W.Ø. Larsen, the family-owned business is now headed by the fifth generation Larsen, Neils. Wilheim Ø. Larsen—the son of the company's founder—was one of the first tobacconists to special order tobaccos such as hand-rolled Russian cigarettes and cigars. Known as the "Old

Man," he established high quality standards that remain ingrained in W.Ø. Larsen's products. In 1913, W.Ø. Larsen was granted the title of "Purveyor to the Royal Danish Court."

In the 1950s, the great grandson, "Ole" Larsen, was dissatisfied with English and French pipes because they were all the same color and style. He wanted a different looking pipe, something that was more creative and exciting. To produce this type of pipe, he commissioned local Danish carvers to create new styles and the Danish-style pipe was born.

W.Ø. Larsen considered mass-producing briar pipes, but the company decided against it almost as quickly as the idea presented itself. Today, Larsen's hand-crafted pipes are created by a team of independent craftsmen who work on their own schedules, carving pipes from their own designs, and using raw materials supplied by Larsen. Larsen believes that this arrange-

*A W.Ø. Larsen horn pipe.
Larsen produces almost 1,000 pipes annually
and one of their better pipes is the horn,
which is comfortable to hold and smoke.*

ment gives the company full control over the quality of wood and does not interfere with artistic endeavors of its carvers.

Pipes with the W.Ø. Larsen name use only Corsican plateau briar. Out of a box of 100 choice blocks, fully one-third will end up in the winter stove. The Larsen craftsmen use bandsaws to cut out the rough shapes of the pipes and sanding discs to give them form. Lathes are used to bore holes and cut stems from vulcanite rods. Handsanding completes the work.[14]

W.Ø. Larsen pipes reflect the attention to detail and quality for which the company is known. The designs are Danish but are no longer "organic" as in the past. Clean lines, beautiful form and balance and great smoking attributes highlight a Larsen pipe. The "Pearl" line of pipes has a superb finish with an unusual whitish-pink tint. Larsen makes almost 1,000 pipes annually but only about thirty Pearls are produced in a year. Retail prices range from $65 to $850.

W.Ø. Larsen also operates a museum highlighting memorabilia from tobacco's rich past. The collection includes an array of antique pipes, tobacco jars, snuffboxes and rare documents.

Nørding

Trained as an engineer, Erik Nørding began pipe carving as a hobby. In the 1960s, Nørding joined some of his fellow Danish carvers and began carving in the freehand style.

Based in Slangerup, Nørding uses superior grade Corsican

Nørding pipes represent a quality pipe with a reasonable price. The Nørding pipe is made of quality briar and the design has a strong Danish tone.

and Grecian briar characterized by its dense grain. He hand cuts mouthpieces from vulcanite bars and crafts inserts and inlays from acrylic and silver. Erik Nørding's pipes are classic Danish. They have a strong conservative tone overlaying the freehand shapes. His pipes are well balanced, have nice proportions and clean lines. Nørding is careful in bringing out the grain in the briar.

In 1995, Nørding launched a series of limited edition pipes, called the "Annual Hunter's Pipe," based on nature's wild animals. Mogens Andersen, one of Denmark's finest illustrators of wild game, draws a different animal for each year, and the drawing is included with the purchase of the pipe. In 1995, the series began with a pheasant pipe. It continued in 1996 with the Sika deer pipe; in 1997, with the Canadian goose pipe; and in 1998, with the beaver pipe. Overall, retail prices of Nørding pipes start at $60 and top $500.

Stanwell

Stanwell is the largest pipe maker in Denmark. Poul Stanwell founded the Copenhagen company in 1942, and today, it produces the finest machine-made pipes in the world. Using fraising machines and very skilled craftsmen, Stanwell makes more than forty different pipe shapes.

Stanwell made the decision a number of years ago to use only Corsican and Grecian plateau briar. The company receives plateau blocks once a year and supervises the wood curing under strict guidelines. The machines are calibrated to make smaller bowls in what has become known as the Stanwell style. The bowls are finished by hand on sanding discs and then hand-sanded. The stems are made from pre-formed Lucite. Stanwell is the cleanest, most organized, pipe manufacturing company in Europe if not the world.

For several years, Stanwell has been using some pipe designs commissioned from top Danish carvers such as Sixten Ivarsson, Lars Ivarsson, Jess Chonowitsch, and Tom Eltang. The strengths of the Stanwell pipe are its high quality plateau briar and the way in which the beauty of the grain is brought out in the pipe. The designs of Stanwell pipes are clean, well proportioned, and balanced. Considering the pipes are machine turned, the number of different and intriguing styles attest to Stanwell's ability and talent. Many of the pipes are very traditional and English in appearance while others are pure Danish in origin, styling, and presentation. Retail prices start at $38 and top out at $275.

France

The Saint-Claude styling of pipes began in the mid-nineteenth century. Inspired by French clay pipes and figural meerschaum pipes from Vienna, Saint-Claude pipes evolved into the classic shapes we are familiar with today. The English, Danes, and Italians then copied the French classic shapes for their own pipes. Italian pipe craftsmen, in particular, would travel to Saint-Claude for employment. When they returned to Italy, they brought with them the latest French technology and styling.

Butz-Choquin

Butz-Choquin—BC—is one of the original briar pipe makers in Saint-Claude. BC was established in 1858; in 1951, the family-owned pipe company of Berrod-Regad acquired Butz-Choquin.

Today, more than 200,000 pipes are made annually by Berrod-Regad, the largest producer of pipes in Saint-Claude. About sixty people are employed making pipes and ten to fifteen percent of the pipes produced are sold in the United States under the BC name.[15] Most of the pipes are made with fraising machines using Moroccan briar ébauchons. The plateau briar is reserved for handmade pipes.

In the 1960s, Claude Berrod began to experiment with pipe styles and introduced the Chateline, Artois, Capitan, and Galion lines.[16] Berrod-Regad also makes pipes for other companies under private labels and employs independent carvers who work outside of the factory. Alain Albuisson is one master carver employed by Berrod-Regad. Influenced by elements from nature, he handcrafts the Collection series for the company. In addition to producing wonderful freehands in the French tradition, he is an exceptional figural carver of busts. He was named "1995 Outstanding Pipemaker of France." Retail prices of BC pipes range from $400 to $1,200.[17]

Butz-Choquin or "BC" is one of the oldest pipe makers in the world and the largest pipe maker in Saint-Claude, France.

Germany

German pipe styles have been influenced by the country's proximity to Denmark, France, and Italy. Many German pipes are classic in nature. The clean lines, smooth, sweeping curves, and wonderful finishes of German freehands and variations on the classic motif reflect a strong Danish influence in particular.

Holger Frickert

Holger Frickert was born in Hamburg in 1949 and acquired a taste for pipe smoking in his teens while on holidays in Denmark. While studying to become a dentist, Frickert worked at a company called Dan Pipe. Here, he met Hans Nielson, Poul Hansen, and Emil Chonowitsch. Each of these carvers shared ideas and techniques with him. Later, he traveled to England and studied there for 18 months.

Frickert believes in crafting pipes that have superb balance and function. He believes in the beauty of a pipe and grain of the wood, but if he must choose between function and the appearance of the grain, he will choose function. His meticulous attention to detail—such as the width of tobacco chamber or

Attention to detail and function, as well as the presentation of the grain are the foundations of Frickert pipes.

thickness of the briar in the shank and bowl walls—attests to his desire to craft a high quality pipe. Some of his work includes the use of dental tools. He even fills flaws like a dentist by scraping or picking out the blemishes.

Today, Frickert is a business partner with Dan Pipe so his carving time is reduced. He has a very organized shop and his tools are arranged and grouped to maximize efficiency. He prefers briar from Greece but uses some Corsican briar.

Frickert first examines the briar block for design possibilities. He then uses a lathe to turn bowls and bore holes while he hand holds the drill bit. His techniques are both English and Danish in nature, especially his preferred methods of staining and curing. Frickert's mouthpieces are hand-shaped from vulcanite rods and the stem accessories include silver bands and horn and acrylic inserts.

Holger Frickert makes a variety of pipes in traditional shapes but with German styling. His pipes are conservative, straightforward—here the English overtones show through—

and well balanced. He refines his classic styles with a Danish interpretation, and several pipe manufacturers such as Stanwell, Ser Jacopo, Bartoli, and Lorenzetti use his designs. Frickert makes a comfortable and pleasant-smoking pipe. Retail prices range from $325 to $700.

Karl-Heinz Joura

Karl-Heinz Joura (pronounced "Your-a") was born in East Germany. At the first opportunity, he and a friend escaped to the west aboard a freighter departing from Rostock. Once free in West Germany, he settled in the Rhineland and found employment with a tool-making factory. Joura had been trained and apprenticed as a tool and dye maker in Rostock. Later, he taught in schools before becoming a full-time carver in Bremen in 1985.

A friend introduced Joura to pipe smoking in the early 1970s, and this experience sparked an interest in pipe carving. In a few short years, he was carving pipes for friends and local shops. Joura has studied techniques of master carvers and worked with some of the Danish masters, but for the most part, he is self-taught.

Joura uses plateau briar from Corsica. Each spring he visits Corsica and hand selects 400 to 500 blocks. When the briar arrives at his shop, he begins the curing process that will last eighteen to twenty-four months. He has also developed an oil

*Many of Joura's pipes have a Danish styling.
He has an interesting way of visually attaching the shank
to the bowl, as if they are two separate entities.*

curing process where the blocks are oiled for ten consecutive days. To begin the 80-step pipe-crafting, Joura sometimes uses a template to draw the outline on the briar block. He drills small guide holes to mark the tobacco chamber and mortise, and turns the bowl on a lathe. As he works the wood, he stops and licks it; the extra moisture brings out the grain for him to follow. After the pipe has been hand sanded, Joura applies the stain. His personal preference is a lighter stain although his pipes come in several different shades. Using vulcanite and ebonite rods from Germany, Joura cuts, carves, and custom fits each stem to insure balance and confront. Joura is very meticulous and his formal technical training insures the pipe has superb craftsmanship and design. He works on each pipe for about a day-and-a-half before completing it and makes about 300 pipes annually.

Joura is known for his very tight, straight-grain pipes. He strives to bring out the heavy density of the birds eye pattern on the bottom and shank area of the pipe. Most of his shapes are classical, but he adds a little wood to the bottom of the bowl. This element, he feels, improves the smoking quality of the pipe and reduces the possibility for burnout.

Joura's pipes have a decided European flavor, reflecting the influence of German, Danish, English, and French designs. Like his Danish counterparts, Joura is influenced by nature—especially the wind, rain, and cold of Northern Europe. His stems and shanks allow him to take some artistic license. As is to be expected, his craftsmanship is Old World, and like so many other Europeans, he lives above his studio. Retail prices range from $750 to $1200 for standard pipes and his masterpiece series, the gold dot pipes, begin at $3,500 and top $7,500.

Manuel Shaabi

Manuel Shaabi was born in Lebanon in 1950. While growing up, he developed a strong passion for wood. The great cedar forests and olive trees of his homeland are some of his most cherished memories. It was only natural, then, that Shaabi decided to make a life in woodcarving. He apprenticed as a wood sculptor and carpenter and worked with some of the finest wood workers in the Middle East.

In his mid-twenties, Shaabi moved to southern Germany and worked as a restoration artist on wooden religious statues in cathedrals. It was during these seven years that his skills and technical abilities were refined. In 1994, Shaabi started carving pipes for Dan Pipe, and in 1998, he founded his own Hamburg shop for pipe making.

Shaabi's pipes include both classic shapes and figural styles. The latter pipes set him apart from other pipe makers. His exquisitely carved figural pipes vary in theme from an octopus and jaguar to Arabic scrolled horns and eagle claws. He uses briar from Corsica, Greece, and Italy and adorns his pipes with silver, gold, amber, and fine hardwood inserts. The classic pipes

Top: Manuel Shaabi's "Octopus" is one of the most unusual creations and is highly sought after.

Bottom: "Ball and talon" and "Arabic scrollwork" pipes. Shaabi is one of the few carvers who ventures into figural pipes.

are roughed out with a bandsaw and then carved and sanded by hand. Because of the ornate designs and features of the figural pipes, Shaabi does all work by hand. His hand tools include chisels, awls, gouges, and knives. Shaabi makes about 200 pipes each year.

Manuel Shaabi makes some incredible pipes. One outstanding example is his disc pipe. The disc is designed vertically on the bowl and the axis is cut slightly off center with the tobacco chamber on one side. This intriguing pipe is a wonder of circles and intersecting lines. One side of the disc exhibits beautiful birdseye grain, and the reverse side has straight grain and birdseye, a rare feat. Retail prices for classic shapes range from $350 to $600 a pipe. The figurals are more expensive, costing between $600 and $3,000.

Shaabi's true talent is brought to light with his disc pipe. The complicated angles and curves speak of his true abilities.

Great Britain

Great Britain has been home to several major pipe manufacturers including Blumfield's Best Briars, Barling, Comoy, and Charatan. The first briar pipe maker in England was Joseph Koppenhagen in 1862. Initially, Great Britain was known more for its clay pipes than wood or meerschaum. This changed as Alfred Dunhill built his pipe business. Dunhill was not impressed with the quality of imported pipes from France or domestically crafted pipes. He built a pipe business that became renowned for its superior craftsmanship and conservative classic designs. The Dunhill pipe established a standard other pipe makers emulate to this day.

Ashton

Ashton Pipes was born in London in 1984, when William John Ashton Taylor left Alfred Dunhill after twenty-five years of service to branch out on his own. It was Ashton's dream to own his own business and to make a superior pipe.

Ashton began making pipes as a teenager and in school he studied engineering. At the age of fourteen, he found a job at Dunhill making cigarette holders. Ashton crafted pipes at

Ashton makes superb natural and sandblast finishes.

Dunhill for eight years and then was promoted to supervisor of the department. After serving eight years as supervisor, he was promoted again to a position overseeing product development, as well as, the quality control of Dunhill tobacco and pipes.

After demonstrating his pipe-making techniques at an industry show, Ashton was contacted by David Field, an avid Dunhill collector and researcher, about the prospects of forming his own pipe company. Arrangements were made to purchase equipment and briar stock. One year later, Ashton Pipes was fully operational.

Finally in business for himself, Ashton was free to experiment with techniques, processes, and designs. Unlike the fraised Dunhill pipes, Ashton wanted to produce a hand-carved pipe that would be "the best smoking pipe in the world." He believed in the oil curing process developed by Dunhill to remove sap from the wood, but felt it needed revising. After much experimentation, he developed a process of soaking the briar blocks in a special vegetable oil bath until completely saturated and then heating them on copper plugs for fourteen days. The heat level was fluctuated to fully force the oil into the wood and remove sap residue. In addition, Ashton was granted a patent in 1988 for a new technique of sandblasting heavy, dense-grained pipes, the first pipe manufacturing patent issued in England in twenty years. The "Pebble Shell" pipe begins with a rusticated surfaced followed by high pressure sandblasting to remove twenty to twenty-five percent of the weight and not damage the grain.

Ashton's approach to pipe making is different from most carvers. He personally selects his Calabrian briar blocks at the sawmill and cures the blocks in England for two to thee years. He uses a lathe to turn the block while he holds a chisel against the turning wood. Using the same machine, he bores the tobacco chamber and the air hole. Sanding discs, hand sanding, and polishing wheels finish the wood in preparation for the custom-made stem. Ashton cuts solid rods of vulcanite into sections and boils them to remove sulfur. Once boiled, the rod becomes harder and the stem resists discoloration. Ashton also uses Plexiglas for stems.

Ashton Pipes employs several people including Ashton's son, Spence, to assist in all aspects of pipe making. Today, the company has expanded into cigars, tobaccos and accessories.[18]

The Ashton pipes are beautiful creations highlighting the natural beauty and grain of the briar. Created in classic shapes of English tradition, the pipes form beautiful straight lines and gentle curving bowls. The coal black stems combined with silver bands and spigots give the pipe a regal appearance. The sandblasted pipes are finished in black with red highlights or dark brown stains matched with brown-colored stems. Ashton pipes are graded by size and stamped by date. Ashton produces between 2,000 and 3,000 pipes annually. Retail prices start at $180 for the sandblast and top at $2,000 for the Magnum.

Castleford

Castleford pipes are named after the many castles that adorn the English landscape; in particular, the castle in Chatham built by the Normans. The Castleford pipe is made by Colin Fromm, a veteran carver of twenty-five years. Fromm was formerly with Ashley pipes before branching out on his own about three years ago.

Based in Chatham, Fromm crafts a traditional English pipe using his own designs. He uses a very high-grade plateau briar from Morocco, Corsica, or Greece. The design is first drawn on a block, and the outline of the pipe is cut on a bandsaw. Fromm uses a lathe to shape the bowl and drill the holes. The precision in which the Castleford pipe is made is a result of Fromm's ability to carefully sight the drill, lathe, and disc with the wood. A sanding disc is used to shape the pipe and Fromm's assistants hand-sand the pipes. Although Fromm often designs and shapes a pipe in about forty-five minutes, the sanding and finishing require much more time. The stems are handcut from superior grade, German vulcanite rods.

Castleford pipes are named after the many castles found in Kent.

The Castleford pipe is the only English-made pipe that is air-dried. The pipes exhibit well-defined grain patterns and most of the shapes are traditional. The few freehands made have a very conservative flavor. Annual production does not exceed 400 pipes. Retail prices range from $70 to $450.

Dunhill

Perhaps the single most recognized symbol in pipedom is the tiny white spot of a Dunhill pipe. Dunhill is a name synonymous with quality. To this day, Dunhill pipes are among the best-made and most expensive pipes in the world.

Alfred Dunhill joined his father's harness-making and carriage-outfitting business in London in 1887 and was given full control in 1893.[19] In 1904, he applied for a patent on his "Dunhill Patent Shield Pipe," a pipe with a raised front on the bowl to prevent ashes from blowing onto the smoker when he was outdoors. In 1907, Dunhill opened his own tobacco shop on Duke Street in

*Shilling, Cumberland, and the Shell Briar pipes.
Dunhill offers a wide variety of pipes with 8 different finishes, 7 colors,
and 25 basic shapes, plus an array of special designs and shapes.*

St. James.[20] Here, Dunhill developed a business philosophy that would guide him and his family for decades.

Dunhill was not impressed with the pipes he was receiving from manufacturers for his store. Determined to offer a better product and increase profits without the expense of a middleman, Dunhill established his own pipe factory in 1910 and employed two master craftsmen from the Charatan factory. In 1912, Dunhill created the famous white spot as an orientation device; it enabled the consumer to correctly reposition the stem after each use. The savvy marketing talents and superb customer-relations skills of Alfred Dunhill helped establish brand loyalty and a customer base that few businesses can claim.

Alfred Dunhill was also talented in other areas of business. He invented the modern pocket cigarette lighter and the sandblasted pipe he termed "Shell Briar." He was the first to use an oil curing process by applying oil to a pipe and then heating it continuously for several days to remove moisture and sap from the briar. This technique provided a lighter pipe and a dry, mellow smoke. Dunhill also branched outside of the tobacco business by adding a line of ink pens, men's fragrances, clothing, and accessories. Cigars, cigarettes, lighters, and other smoking accessories were also added to the Dunhill line. In 1924, Alfred Dunhill wrote *The Pipe Book*, a landmark study still used as a reference work today.

Dunhill pipes are crafted by hand with utmost precision. He uses traditional shapes that are turned and finished by hand. Dunhill bowls are smaller and usually have a darker finish, giving the customer a smaller, less noticeable pipe, but one styled in a manner to match his conservative English clothing. The darker bowl brings less attention to the smoker and helps downplay the appearance of a new pipe.

Only one or two out of one hundred to two hundred premium quality briar blocks gives rise to a Dunhill pipe. The Dunhill people examine thousands of blocks in search of the few hundred that are flawless. The demand for superior quality blocks is so staggering that Dunhill has begun to purchase ébauchons from other pipe makers that meet the Company's specifications. The pipes are finished with stains, waxes, and hand-cut mouthpieces from vulcanite rods. Each stem is fitted perfectly, the white spot is added, and then the pipe is sent through rigorous inspections before the name "Dunhill" is stamped on it. Neither flaws nor putty fills are accepted on Dunhill pipes, and the company sells the pipes failing the tests, known as "fallings," under another name.[21]

The Dunhill pipe has become one of the most collected pipes in history. Collectors of antique pipes pursue with diligence older Dunhills. The majority of Dunhill pipes are sandblasted and many different finishes and sizes are available. "The Bruyere," introduced in 1910, has a smooth finish with a deep red or mahogany color and black vulcanite stem and is still available today. The highest quality straight-grain Dunhill pipe is the "Root Briar." Introduced in 1931, it has a smooth finish with a golden color stain to bring out the beauty of the superb straight grain.

The Dunhill philosophy—"It must be useful. It must work dependably. It must be beautiful. It must last. It must be the best of its kind."— still guides the Dunhill Company today.[22] Retail prices begin at $310 for a Shell Briar and reach $3,185 for a straight grain.

Ferndown

Ferndown is a high-quality briar pipe made by Les and Dolly Wood. The Ferndown pipe was born in 1983 in London when Les, a silversmith at Dunhill, left Dunhill to build his own company. He chose the name Ferndown in honor of his home.

Ferndown pipes are exceptionally well made in the old Dunhill tradition. And like the Dunhills, the pipes are oil cured.

In addition, the Ferndowns have some of the finest silverwork of any pipe made thanks to Les's skills. Dolly hand finishes the pipes. Ferndown pipes are very traditional in shape and size and have a very English appearance. Built for balance, the pipes are streamline and have clean lines. The pipes have a nutty flavor from the start when smoked and each pipe is stamped "L & J. S." The pipes are available in several finishes, including black, brown, unstained, and deep plum. Ferndown produces about 2,000 pipes annually. Prices range from $195 to $600.

Ferndown pipes are noted for their conservative English styling and fabulous silver work. They are easily recognizable by the LJS stamp on the pipe.

Ireland

Pipe making in Ireland is largely a cottage industry. Pipe shapes have changed little over the years. In general, Irish pipes are small and classic, with a full-bent shape.

Peterson

Peterson of Dublin is one of the oldest names in briar pipes. Peterson began as a small tobacco shop in 1865. Friedrich and Heinrich Kapp, opened Kapp Brothers as a place to purchase meerschaum and briarroot pipes. Charles Peterson, a pipe maker, approached the brothers with a new pipe design and a plan to form a partnership. The brothers agreed and Kapp & Peterson was born.

Kapp & Peterson became Dublin's most respected pipe and tobacco manufacturer. In 1890, Peterson invented the now-famous Peterson's Patented Smoking System pipe. This pipe was designed to contain an interior secondary bowl or reservoir in the shank, trapping condensed tobacco moisture from the smoke and giving the smoker a cooler and drier smoke. In 1898, Peterson invented the Peterson Lip mouthpiece, which directed the smoke up and away from the tongue. This pipe reduced the amount of salvia entering the stem and provided a contoured fit for the mouth.

Today, the many styles and shapes of Peterson pipes are very traditional in nature and are wonderful representations of pipes from the nineteenth century. Each year, Peterson presents a limited edition set of pipes crafted by their finest master craftsmen. Retail prices range from $53 to at $650.

The "Emerald" series has both rusticated and smooth finishes and is but one of many styles Peterson offers.

The Watson pipe is one of seven shapes in the original Sherlock Holmes series.

Italy

Before World War II, Italian pipes were known for their quantity rather than their quality. The large Rossi plant in Northern Italy, for example, produced a boxcar of pipes a day. After World War II, Italian pipe makers such as Savinelli, Brebbia, and Castello worked to produce a quality product with pleasing designs.

Today, Italy has two schools of pipe design and manufacture. The oldest is in Northern Italy where companies like Savinelli, Castello, Brebbia, Ascorti, Radice, and Ardor are located. Using machines to turn their bowls, carvers of this school produce classic pipe shapes.[23]

The second school of pipe making in Italy is known as the Pesaro or Baroque school. Based in Pesaro on the Adriatic Sea, members of this school include its founders Giancarlo Guidi of Ser Jacopo, Mastro de Paja, Don Carlos, and Il Ceppo. These carvers experiment with classic design and ornamentation, borrowing concepts and designs from Danish artisans.[24]

In Italy, the pipe making trade is often passed down from generation to generation and whole families may work in one factory or shop. Many of Italy's pipe craftspeople do move from one manufacturer or shop to another, however.

One of the more unique pipe lines from Ardor is the RoverArt. This line showcases combined artistic displays of rustication and natural smooth finishes.

Ardor

Located in Gavirate, Ardor is one of the older pipe manufacturers in Italy. Beginning in 1911, the Rovera brothers—Federico, Carlo, Cornelio, and Francesco—joined together to create a pipe making business under the Rovera banner. Federico's son, Angelo, joined the company in the mid-1940s and learned the art of pipe making. In 1958, Angelo Rovera brought in his son, Dorelio, as well.

For several years, Angelo and Dorelio Rovera perfected their technique and design. The Ardor pipe grew from this union. The name Ardor was derived by using each of their names, "Ar" for Angelo Rovera and "dor" for Dorelio Rovera. Today, Dorelio is the administrator of the Ardor Pipe Company.

Ardor's Italian plateau briar from Sardinia, Liguria, and Calabria is personally chosen by Dorelio. The briar blocks are hand cured for ten years before being hand carved into pipes. In the early years of the Rovera pipes, water mills were used to operate pipe-making machines. Today, all Ardor pipes destined for the American market are made by hand using lathes. The mouthpieces are handcut from acrylic slabs.

Ardor makes three lines of pipes. The Ardor Pipe is available in nine pipe models including the Pipe Scolpite—figural pipes displaying faces of famous people and animals. The Rover Art series possess a rusticated finish superimposed with smooth geometric and scrollwork carvings. The third line is the Varese.

Ardor makes beautiful pipes. They are meticulously finished, with finely executed sandblasting. Bamboo is now used as a shank material and the silver work of the Ardor craftsmen is creative and in balance with the pipe. Currently, Ardor makes available more than 125 shapes, ranging from the classical to the highly unusual. Retail prices range from $160 to $850.

Paolo Becker and Becker & Musicò

Paolo Becker of Rome is a prolific and talented carver. His talent is represented in two boutique workshops: Paolo Becker and Becker & Musicò. His father, Fritz Becker, a pipe smoker for fifty years, began making pipes as a hobby in 1974. Becker, also interested in woodcarving and design, joined his father in 1977.

Today, Becker makes about two hundred hand-carved pipes a year under the name of Paolo Becker. His goal is to make a pipe that offers a "perfect smoke and good taste." He hand-picks his briar from the sawmills twice a year, and cures it an additional three years after receiving it to ensure a very dry smoke. Becker carves by sight changing the shape of the pipe to compensate for the flaws in the wood. He roughs out the pipe from the Italian briar with a sanding disc and then bores the holes with a drill. Becker does his sanding by hand and uses steel wool to give the pipe its final finish. He hand carves each stem to match the pipe from vulcanite or Plexiglas.

Becker says "each pipe is my own design and reflects my personality." His pipes come in both classic shapes and conservative freehands. He is very particular with the look of the wood and how the pipe smokes. Paolo Becker works alone in his shop when he makes Paolo Becker pipes. His retail prices are $350 to $2,000.

When Paolo isn't carving Paolo Becker pipes, he can be found working with Massimo Musicò. Together they produce very traditional, English style pipes. These pipes are noted for

Becker hand carves only about 200 pipes each year. His style is classic or conservative freehand with a hint of Danish overtone.

their clean lines, well defined shapes, and balanced proportions. The pipes are very striking and resemble the Dunhills.

For these traditional pipes, Becker turns Calabrian briar on machines and finishes the bowls by hand. Musicò fits the cast stock vulcanite mouthpieces to the pipes and crafts silver bands and spigots. Most of the pipes are sandblasted and finished in black or tan. Total annual production is around 2,000 pipes. Retail prices range from $135 to $320.

Brebbia

Brebbia is located in the small village of Brebbia between two large lakes and in the shadow of the Swiss Alps of northwest Italy. In 1947, cousins Enea Buzzi and Achille Savinelli opened a pipe manufacturing facility in the building housing the power plant. This provided free electricity to run saws and other equipment needed to produce pipes. In just a few years, this union of the cousins would help change the pipe industry in Italy.[25]

Achille Savinelli oversaw the distribution and sales of the pipes from the family-owned Savinelli Company headquarters in Milan, and Enea Buzzi managed the pipe factory in Brebbia. The new line of pipes carried the Savinelli name. The philosophy of this new company was to produce quality pipes, an unfamiliar concept in war-ravaged Italy. Six years after the company was founded, differences of opinion caused a rift in the company. Buzzi began manufacturing pipes for other tobacco

Brebbia makes six different lines of pipes with about 70 shapes and a variety of finishes.

shops and changed the name of his company to Pipe Brebbia. Achille Savinelli opened his own pipe factory in Barasso in 1956.[26] Buzzi's son, Luciano, came aboard in 1977.

Today, Pipe Brebbia employs seventeen people of whom fourteen make pipes, two specializing in freehands. Pipe Brebbia uses Ligurian Italian briar. When the blocks arrive, they are stored in a high humidity and high temperature climate. The wood is slowly dried for four to five years before it is ready to be carved into a pipe. After the pipe has been shaped by a fraising machine and finished by hand, it is set aside for four to six months for additional drying. Luciano Buzzi believes this improves the smoking abilities of the pipe.[27] Pipe Brebbia was one of first manufactures to use acrylic mouthpieces, and today all mouthpieces are made by hand.

The majority of the 14,000 pipes Brebbia manufactures annually are the less expensive, machine-made pipes using ébauchon blocks. Brebbia's high-grade, handmade pipes are carved from plateau briar. The artisans carefully work the wood to bring out the very best in the straight-grain briar. Brebbia only makes about 1,000 straight-grained pipes per year. Retail prices of all of the pipes range from $45 to $350.

In addition to producing beautiful pipes, Pipe Brebbia has its own museum of antique pipes from around the world. The company collects more than 6,000 pipes.

Castello

Castello, the Rolls Royce of Italian pipe companies, carries on the pipe making tradition established on May 17, 1947, when Carlo Scotti and his team made their first pipe after seven months of carving and experimentation. In 1950, they introduced the acrylic mouthpiece to the world. Today, Carlo Scotti's son-in-law, Franco Coppo, directs the Castello operation from Cantú.

Before a Castello pipe is carved, it is designed according to time-proven ratios, dimensions, and shapes to bring out the very best of the highest-grade briar block. The Castello tradition then entails shaping the bowl followed by drilling the shank and bowl chambers. Each pipe is handmade, and remarkably, each of the six Castello artisans is trained to perform all of the pipe-making functions. Only forty percent of the 6,000 Castello pipes made annually are shipped abroad, the rest remain in Italy.

The Castello pipes are carved from the finest briar. Castello briar has been seasoned for at least six years and sometimes as much as nine years, a standard most companies cannot claim. The Castello philosophy for curing briar is to store the blocks in the workshop where the wood undergoes seasonal changes and the daily heating and cooling cycle. The Castello philosophy also means, "Knowing how to work at the right moment, not before or after. Knowing how to work with respect for precise parameters of functionalism, aesthetics and good taste."

The lines of a Castello pipe are slick and elegant. The straight grain is impeccable as is the balance. Most of the Castello pipes are classic in design and very conservative in nature. Even the freestyle Castellos are very conservative. The

Pipes from Castello have an elegant appearance.
The wood is only the finest and is graded by skilled workers
to produce only the best pipes.

medium size of the Castello classic pipes also remains true to Italian traditions.

Castellos come in twelve series and are graded by size in most of the series. The "Sea Rock" and "Old Antiquari" series are priced in the low $200 range while the "Collection Fiammata" and "Collection Greatline" series peak at $1,300 and $2,900 respectively.

Il Ceppo

In 1978, Giorgio Imperatori of Pesaro began making a new pipe line named Il Ceppo, meaning "the root." In 1993, Franco Rossi joined Imperatori bringing with him a style reflecting the Mastro de Paja and Ser Jacopo traditions. Together the two carvers have combined talents, techniques, and creativity to produce a sophisticated pipe with a neo-classic look. Imperatori has since retired, but his wife, Nadia, still works with her brother, Franco Rossi, and a new carver, Massimo Palazzi.[28]

Il Ceppo uses Calabrian briar for their pipes. The two carvers design their pipes by first examining the block of wood. Once the style has been decided upon, the shape is drawn out on the block. The blocks are carved by hand and hand sanded. The stems are hand-cut from sheets of Plexiglas.

The classically shape pipes reflect Imperatori's influence and training from the early Mastro de Paja school. Rossi's freehands, in contrast, have more of Ser Jacopo feel. Il Ceppo pipes

In addition to creative designs, Il Ceppo is careful in presenting the beauty and quality of the straight Calabrian briar.

are well balanced, have clean lines, and are well proportioned. The stem design is very creative and reinforces the beauty of the pipes. Their use of silver ornamentation is superb. Annual production is between 2,000 and 3,000 pipes. Retail prices are from $150 to $600.

Mastro de Paja

Mastro de Paja (meaning "master of the straw") was established in 1972 in Pesaro by Giancarlo Guidi. Guidi, a self-taught pipe carver,[29] began making pipes while attending the university studying art and design. He was so passionate about carving pipes that he decided to make pipe carving his vocation. His first tools were a lathe, saw and a sander.[30]

Guidi left Mastro in 1982 to form Ser Jacopo. Following his departure, Mastro went through many changes and several business partners. In 1982, Alberto Montini was brought on board

The Mastro de Paja line of pipes offers many shapes, styles, and finishes in the classic and freehand designs.

as general manager. Thomas Cristiano, joined the Company in 1990 to promote Mastro in the United States and four years later, Cristiano became a partner. In 1997, Cristiano left Mastro, and returned full time to his company, Cristom.[31]

Today, the shop employs six full-time carvers with an average of ten years of carving experience. Out of the six artisans, four have degrees in art from the University of Urbino's Academy of Fine Arts and two have expertise in fine metal arts. In the last several years, master carvers Vittorio del Vecchio and Mario Pascucci have introduced new styles in the Company's lineup. Mastro produces freehands, classics, and classics with a contemporary flair.

Most of the Mastro pipes are made by hand, although some are made with fraising machines using templates for classic shapes. For all Mastro pipes, the tobacco chamber, mortise hole, and air hole are each drilled separately and finished by hand.

Mastro's craftsmen use plateau briar from Corsica or Calabria and begin the pipe with a design sketched on the block. The wood is cut first with a saw and then shaped with sanders. Once final sanding has been done by hand, it is fitted with an acrylic or cumberland stem.

Mastro offers a variety of pipe designs from the very classic to freehands with conservative Italian overtones. Some of the freehands have intriguing shapes that testify to the skill and talent of its artistically trained staff. One of the most beautiful pipes made by Mastro is the large ball pipe. About the size of a tennis ball, the bowl is perfectly round, a task not easily accomplished.

Mastro does an excellent job in revealing the beauty of the grain in their pipes especially in the high-end pieces. The lighter finishes used on some Mastro pipes give the grain the appearance of fire. Mastro also has a variegated brown finish that is most appealing. The metalwork on Mastro pipes is superb and intriguing as well. Mastro makes about 2,000 pipes annually. They retail from $150 to $1,000.

Radice

Before Luigi "Gigi" Radice developed his own line of pipes in 1981, he worked for Castello. In 1969, he and Pepino Ascorti, of Ascorti Pipes, left Castello to form their own pipe company, Caminetto. When Radice left Caminetto in 1981, he formed Radice pipes.

The Radice horn pipe is one of the most difficult pipes to make. This pipe illustrates exquisite detailing in the horn and briar sections.

Radice's two sons, Marzio and Gian Luca, have been masterfully trained by him, and today, the three Radices work in their large shop behind their home in Cucciago. Using Ligurian plateau briar, the family hand carves classic-shaped pipes in the traditional manner. They each begin with a design in mind and carve the pipe until the idea becomes a reality. The pipes are shaped by hand with grinding discs or turned by machine, then sanded by hand and custom fitted with hand-cut black Plexiglas stems.

The talent and skills of the Radice family are evident in their beautifully carved pipes. There is a certain flair in the Radice pipe, and the Radices continue to explore new artistic designs and styling. In addition to using well-aged bamboo for inserts, for example, the artists have developed a handsome look by carving the briar shanks and inserts to resemble bamboo. Another Radice presentation is the use of horn lined with briar. The Radice pipes are also known for their excellent use of silver. Another effective technique introduced by Radice is the "silk cut" sandblast finish that has a fine texture similar to an orange peel.

Radice makes more than 3,000 pipes annually. Retail prices range from $195 to $500.

Savinelli

The Savinelli family name has a long history in the world of pipes. Achille Savinelli began selling pipes and accessories in his Orefici Street Shop in Milan in 1876. Later Achille began designing pipes and he employed local talent to carve the pipes. His son Carlos entered the business and his grandson, Achille, eventually joined the company as well.

In 1948, the young Achille Savinelli opened a pipe factory near Brebbia, with relative Enea Buzzi. But this union was short lived. In 1956, Savinelli opened another pipe factory in Barasso, Italy. It is here that the name Savinelli became synonymous with fine pipe making. In fact, the Italian pipe indu stry as a whole improved in quality and gained respect because of Savinelli's influence. In addition to running the family company, Achille Savinelli continued to design and create new styles of Savinelli pipes at his own work bench.

When Achille Savinelli died in 1987, his daughter, Marisetta, and son, Giancarlo, took over the company as equal owners. Trained in the field of fashion design, Marisetta developed acces-

*This pipe was made to commemorate
the 120th anniversary of the Savinelli company.*

sories marketed by Savinelli including cigar accessories, lighters, men's fragrances, and shampoos. Giancarlo Savinelli has assumed managing the daily operations of the company and has designed most of the new pipes since his father's passing.

Savinelli uses briar from Italy, Corsica, and Sardinia. The high quality briar is cured in the traditional manner by drying naturally in open-air buildings for at least two years. The handcrafted pipes at Savinelli begin as designs drawn on blocks and then are roughly shaped with saws. The pipes are then handshaped on sanding wheels by craftsmen and finished with natural stains, oils, and canuaba wax to give the pipes their Savinelli look.

Savinelli's production line is divided into two sections, the pipes made entirely by hand (the Autograph, Linea Piu', Artisan, and Collection series) and the pipes shaped by machines and finished by hand (the Classic series). Only one pipe in a thousand will become a Giubileo d'Oro, the perfect pipe. Amedeo Bogni, a master artisan with fifty years experience of which the last thirteen have been with Savinelli, carves most of the Autograph series. Ignazio Guarino, head of the shaping department, and Franco Toscan have thirty and thirty-six years of experience respectfully, apply the finishing touches to the Autograph pipes.[32]

Savinelli pipes are well balanced, have clean lines and represent the finer points of the Italian pipe tradition. The Autograph freehands are works of art, sculpted with the utmost care and sensitively to elegance and design. The lines of the pipe flow gracefully around the bowl enticing the smoker to hold and admire its natural-looking beauty.

Today, Savinelli makes seventy-two different pipe shapes each available in different finishes. It is one of the largest pipe makers in the world and each year they produce over 100,000 pipes. The Savinelli factory also makes additional pipes as con-

The "Autograph" series represents the top of the line Savinelli pipes.

tract work for other brand names. Prices range from $45 to $1,100. Only forty percent of the pipes made by Savinelli carry the company symbol, the other sixty-percent are sold under other names around the world.[33]

Ser Jacopo

Giancarlo Guidi is the heart and soul of Ser Jacopo. A master carver, designer, and artist, he's made the Ser Jacopo pipe unlike any other Italian pipe. He named his company after his young son who died early in life, "Sir Jacopo."

In 1982, after Guidi left Mastro de Paja, he returned to his roots and reopened the old shop behind his parent's house in Pesaro. His old company, Mastro, had grown into a large organization of about thirty employees and Guidi missed working alone or with a small team. His old friend, Bruto Sordini, also left Mastro to join him as a partner at Ser Jacopo. The Ser Jacopo pipe instantly became a work of art. After six years with Ser Jacopo, Sordini left to form his own pipe company, Don Carlos. Today, Giancarlo has a staff of six full-time and two part-time employees.

From the start, Ser Jacopo pipes have had a distinguished shape and their orange color quickly became a recognizable symbol. Guidi buys his plateau briar direct from sawmills in Liguria and Corsica. He designs and cuts the shapes of his pipes with a bandsaw. His team drills the holes on a lathe and shapes the pipes with sanding discs followed by much hand sanding and buffing. Each stem is handcut from square Plexiglas rods. Ser Jacopo is known throughout the world for its superior designed and crafted stems.

Although Ser Jacopo pipes have a very traditional English appearance, the company often incorporates innovative design and construction techniques. This is especially true of the freehands, many of which are intriguing because Guidi enjoys experimenting with different styles and pipe designs from the past. It is as if he pulls parts from several pipes to give a nontraditional pipe traditional parts. One of his more successful programs is the Picta series. This edition includes sixteen briar pipe shapes inspired by pipes portrayed in Van Gogh's paintings. A color pamphlet on the series accompanies each pipe. Guidi's newest creation is a two-pipe series based on the American Indian peace pipe. Today, Ser Jacopo makes between 5,000 and 6,000 pipes. They retail from $195 to $5,500.

This pipe is one of sixteen shapes from Ser Jacopo's "Picta" series. Each pipe is derived from a pipe in a Van Gogh painting.

Japan

Pipes from Japan reflect several design influences. Of particular importance is the Danish penchant for clean lines, simple elegance, and smallness in size. The Japanese marriage of bamboo and briar is a masterful stroke and some of the bowls and stems resemble the kiseru (pipes) of earlier times.

Shizuo Arita

Shizuo Arita is a retired company executive who carves pipes as a pastime. In 1975, Arita received a pipe carving kit from his wife as a birthday gift, and two years later, he won first place in a pipe-carving contest. Buoyed by this success, he began selling his pipes at Kagaya Shop in Tokyo.

A self-taught artisan, Arita has learned carving by trial and error. He designs his pipes to bring out the best in the grain and provides visual interest with flowing curves and intriguing stem designs. His shapes are classical with a bit of a Danish feel.

Arita begins a pipe by drawing a sketch and then matches the design to a block of briar. He cuts out the outline of the pipe with a bandsaw and shapes the pipe on a lathe and with metal files. He does not use an electric sander. He feels that the use of a

Arita's pipes have a Danish appearance. From the lines and curves of the shank and the bowl to the final finish, they reflect nuances of Danish makers.

metal file is the only way to create the gentle curves found in his pipes. He uses a variety of materials as inserts and rings on his pipes including water buffalo horn, staghorn, whale teeth, coral, shells, bamboo roots, boxwood, ebony, mahogany, rosewood, maple, and stainless steel.

Arita "designs the pipe to the taste of the smoker." Arita makes between 70 and 100 pipes each year and they retail from $250 to $850.

Jun'ichiro Higuchi

Jun'ichiro Higuchi began carving pipes in 1974 when he purchased a pipe kit. He had always admired woodcarving, and during the 1970s he visited with other pipe carvers in coffee shops in Tokyo to exchange ideas.

Today, Higuchi makes about fifty pipes a year. His designs incorporate a natural harmony and balance. Once he decides on a design, he finds a piece of briar that will yield the pipe he wants. He follows the grain as he shapes the pipe on a lathe. He bores tobacco and air holes and then refines the shape with a belt sander, files, and carving knives he has designed himself. His stems are hand-crafted from ebonite and water oxen horn, and he crafts inserts from ivory, horn, and boxwood.

Higuchi makes pipes that have a conservative Danish appearance and resemble S. Bang and Lars Ivarsson pipes. In each pipe, he beautifully presents the straight grain of his plateau briar, again very much like the Danish masters. Average retail prices start at $250 and can reach up to $600.

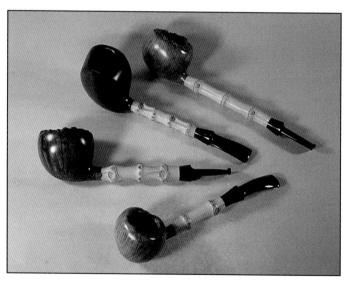

Higuchi gracefully combines the flavor of Eastern and Western cultures into his pipes. His use of bamboo stems and small bowls reiterates the feeling of earlier Japanese pipes or kiseru.

Tsuge Pipe Company

For at least two centuries, the Tsuges has been a family of highly skilled and elite craftsmen. Originally, the Tsuge family fashioned swords for samurai and their craft earned them status in Japanese society.

Kyoichiro Tsuge, head of the Tsuge Company, was born in 1910. His father managed the cigarette plant of Towa Tobacco Company in Seoul, Korea. At age thirteen, Kyoichiro was orphaned. He returned to Japan and lived with an uncle. He became an apprentice at a cigarette holder manufacturer where he learned to craft ivory cigarette holders. When Tsuge turned twenty-six, he married and started his own cigarette holder making business, specializing in ivory. After World War II, Tsuge returned to Tokyo and resumed his business. Since Japan had no cigarettes, he began making pipes. Briarwood was not available in Japan following the war, but high-quality cherrywood was common. His company also made ivory souvenirs and tobacco accessories.

Today, the Tsuge Company has branched out and Tsuge's three sons assist him in running the company. Tsuge has two factories that produce machine-made pipes and a workshop that produces the "Ikebana" line of very fine hand-crafted pipes. Tsuge pipe production reached its peak in the mid-1970s before the price of the yen began climbing.

The Tsuge handmade pipe is meticulously crafted. Tra-

The simplistic forms of these Tsuge pipes add to the serene beauty and superb craftsmanship found in the Ikebana line.

ditional means are still practiced but the machinery has changed a little, as a computer-driven lathe turns the Tsuge bowl with an accuracy of two one-thousandths of a millimeter. Only one to two pipes are made each day.

Tsuge makes very traditional classic pipes and freehands with a Danish appearance. The pipes are available in a variety of shapes and finishes (including sandblast), and many have bamboo inserts. Both the classic shapes and freehands are beautifully made and balanced. The briar is carved to bring out the best in the grain. To glance at the pipes, you would think S. Bang, Lars Ivarsson or Jess Chonowitsch carved them. Retail prices for the Ikebana line are $200 to $8,400.

Spain

Like the German and Italian pipes, Spanish pipe styles are governed by time-honored traditions and reflect stylistic influences from nearby pipe-making centers.

Joan Saladich y Garriga

The pipes carved by Joan Saladich y Garriga of Sabadell are works of art. This self-taught individual carves pipes like the Viennese carvers of the last century with the exception that Saladich does his work in briarwood.

Saladich began smoking a pipe at age sixteen and not long afterward began collecting pipes. At the age of twenty, he began carving pipes for himself. He also did pipe repairs, which taught him much about pipe construction and carving techniques. The pipes he carved initially were freehand in style followed later by classic shapes. Once he saw the magnificent meerschaum pipes from Vienna, his tastes in pipes changed and he began carving figural pipes. Saladich has carved pipes from a variety of materials including meerschaum, cherry, tuya, ivory, amber, pressed amber, bone, acrylic, and briar.

Today, Saladich uses briar for his pipes. He personally

Saladich is one of only a handful of artists in the world who carves figural pipes. His magnificent works of art are reminiscent of pipe artisans of the 1800s.

Saladich's work is so highly detailed and beautifully executed that it requires 30 to 80 hours of his time for each pipe.

selects his blocks of briar from an ancient sawmill and slowly air-dries the boiled wood. Dense and compact briar is used in figural pipes and straight grain briar is reserved for classic-shaped pipes.

All of Saladich's pipes are made by hand. The only power tools he uses are a saw, lathe, a Dremel-like tool with screws and mandrels for detail work, and polishing machines. His hand tools include chisels, gouges, knives, rasps, and sandpaper. He begins a new pipe by drawing sketches on paper to create a design, and finishes a pipe with hand-crafted acrylic stems. (Some of his figural pipes have horn or amber stems.) He also uses gold, silver, ivory, and amber as inserts. Because his work is so detailed and the briar is very hard, his busts require about thirty hours of work and full-length figures on pipes need about eighty hours of work.

Saladich creates beautiful classic and figural pipes that resemble early twentieth-century shapes more than contemporary classics as seen in Italy and the rest of Europe. His stems complement the pipes and he uses larger shanks as compared to other carvers. Saladich's artistic talent is applied so gracefully to wood that his pipes have a finesse few carvers can achieve. His artistic abilities can be compared with the old Viennese carvers, and Joan's unique talents in figurals may be the finest in the world. Retail prices begin at $350 for classic shapes and $650 for figurals.

Sweden

Pipes from Sweden have a distinct Scandinavian appearance akin to its Danish neighbor to the south. Bo Nordh's influence in the pipe world in particular has placed Sweden at the forefront of design and technique in modern briar pipe crafting. His work combined with the artistry of S. Bang, Jess Chonowitsch, and Lars Ivarsson has redefined the modern briar pipe.

Bo Nordh

Based in Malmo, Bo Nordh is one of the finest pipe makers in the world today. Each of his pipes is a masterpiece and he only makes about forty to fifty of them each year.

Nordh began smoking pipes at an early age. After finishing college with an engineering degree in machine technology, his wife, Birgit, suggested he visit the local tobacconist and try his hand at carving a pipe from a starter kit. Being a perfectionist, Bo was never quite happy with the kits. However, he exhibited talent and his pipes sold. Urged by his tobacconist, Olle Johnson, Nordh traveled to Copenhagen and met Sixten Ivarsson. Sixten shared his pipe making knowledge and technique with Bo.

Nordh treats his briar with the utmost respect. He pays up to

$100 a block for the very best briar available, but for every pipe he carves, he discards a block of briar because of his extremely high expectations he has of the wood. When a shipment of Corsican blocks arrives, he cleans the rough outer part of the briar with a steel brush and roughs the smooth cuts with a grinder. Then the briar goes into storage for slow drying. When the humidity and temperature in his shop are consistent with the storage area, he moves some of the briar blocks into the shop. There they will remain for up to five years, waiting for the moment Nordh decides it is time for the block to be fashioned into a pipe.[34]

Nordh examines each block in search of the perfect specimen for a pipe. He studies the grain for imperfections and areas that may require careful carving. He pencils the pipe shape on a block and then carves the briar by hand on a grinding disk. The surface is smoothed with a hand file. Additional shaping is done with a knife or differently sized grinding stones. The file is once again called into action to smooth and finish the wood. Once the wood has been carved, smoothed, and sanded, Nordh guides the drill press to bore the tobacco chamber, mortise, and air hole. Calipers are used to check the shape of the pipe and the thickness of the bowl's interior wall. The pipe is then stained using a pipe cleaner and set aside for a couple of months until the humidity is right before the stem is fitted.

Most of Nordh's pipes are made with ebonite stems and hand-finished to match the shank. Occasionally he uses bamboo or staghorn as inserts between the shank and the stem. His pipes reflect his understanding of balance, and he has developed a great respect for lightweight, superior smoking pipes. Bo Nordh signs his pipes with a metal stamp: B. Nordh, Sweden.

Nordh, like other Scandinavian carvers, bases much of his design and form on nature. His pipes are works of art that presents a sense of movement with wonderful sweeping designs. The flowing curves and delicate angles resemble shells, horns, and waves. The majority of Nordh's pipes are freehands. His trademark styles include a horn, the ballerina, and the pipe he is most noted for, the shell pipe. This exquisite masterpiece is shaped like a conical shell. The spiraling curves of the pipe are truly astonishing, created with utmost precision and balance few carvers in the world can match. He has made only five shell pipes in forty years and when one is available, it starts at $6,000. If you are looking for one of the finest crafted and smokeable pipes in the world today, you should purchase Bo Nordh. Retail prices on his pipes begin at $1,800 for a sandblast and go up, when one is available.

One of the most stunning achievements in pipe design and balance is found in this ballerina pipe by Bo Nordh.

Turkey

Meerschaum carving in Turkey is performed differently than pipe carving in other countries. Turkey's carving society is very homogenized, and carving methods are similar. Instead of carving or turning a pipe at a table like briar artisans, meerschaum carvers in Turkey carve in a seated position, oftentimes sitting on one leg turned underneath and using the thigh and knee of the free leg as a workbench. Some artisans prefer working alone and others work in teams. Like so many other pipe carvers around the world, many of the Turkish carvers work out of their homes.

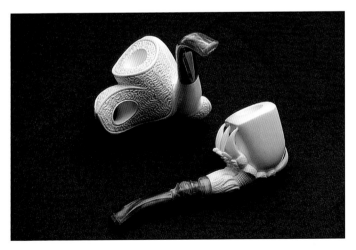

These two pipes exemplify the varied carving style of Ismet Beckler. Both the abstract and the stylized claw are works of art.

Ismet Bekler

Ismet Bekler is probably the most recognized pipe carver in Turkey. Bekler makes between 2,000 and 4,000 pipes annually and his signature series includes up to 500 pipes each year. His work ranges from abstract to figural and classic shapes to ornate embellishments. One of his more popular pipes is the Laughing Bacchas and requires two days to complete. He applies scroll and patchwork ornamentation to the bowl and rim. Retail prices range from $15 to $500 for normal pipes, and up to $35,000 for a collectors series.

Yunus' work focuses on realistic detail of mystical characters and personages.

Yunas Ege

Ege carves faces of people and animals into his pipes. His work is very realistic and well crafted. He makes about 500 pipes each year.

Sevket Gezer

Gezer crafts several different styles of pipes. His calabash pipe, for instance, is heavily ornamented with floral and scroll work and geometric bands. Another style of pipe features a panel on each side encircling an animal such as a dragon, eagle, or lion. The decoration between the panels is a sweeping floral design that reinforces the circular pattern of the round panels. He also carves animal heads.

Huseyin and Mustafa Sekircioglu

The Sekircioglu brothers carve mostly classic meerschaum pipes. The only animal they carve is an eagle claw. The Sekircioglus are devout Muslims and it is their belief that carving animal faces is sacrilegious. Using a lathe they produce beautiful billiards, bents, bulldogs, Canadians, and calabash pipes. Their annual production numbers in the low thousands.

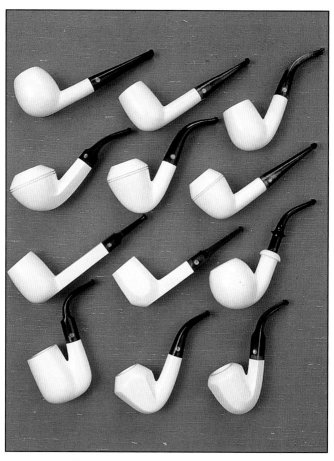

The Sekircioglu brothers produce a wide variety of pipes including ornately decorated and nondescript designs.

Salim Sener

Sener's pipes reflect a low-relief carving style, with baroque features including scrollwork and floral designs. His "signature" pieces include a reverse eagle claw and an eagle claw holding a rose.

Sadik Yanik

Yanik's style is different from other Turkish carvers. He makes reproductions of meerschaum pipes of one hundred years ago. His work includes an assortment of animals, mythical characters, battle scenes, and events from everyday life. Yanik is a young carver in his mid-thirties and has been carving for about fifteen years. His father, Huseyin Yanik trained him, and his talent offers much promise for the future.

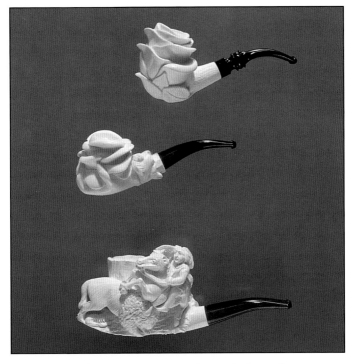

Yanik's presentation in pipes is very similar to the meerschaum pipe styles of the late nineteenth and early twentieth centuries. Many of his characters are from mythological stories.

United States

In the United States many pipe styles abound. American pipes reflect Danish, Italian, English, German, and French influences. Often, pipe makers who live only a few miles apart carve pipes very differently. American pipe makers exhibit an individual spirit not unlike the one that molded the country. As you read the individual sections on the American carvers, you will get a sense of their numerous techniques and styles. I think you will agree that America is really a great melting pot for pipe making. The American pipe artisans reflect a sense of variety, individuality, and accomplishment.

E. Andrew, Briars

Living in Milwaukee, Wisconsin, Edward Andrew Jurkiewicz happened into pipe making by accident. He had been smoking pipes, particularly his father's pipes. In 1975, he answered an ad from Amphora tobacco for a Danish freehand costing only $15 plus two empty packs of Amphora. He found that the pipe was wonderful, so he ordered another Amphora pipe, a large freehand with a shape he was less than fond of. After showing the pipe to Jack Uhle at his tobacco shop, Uhle challenged him to

Andrews makes about 150 pipes annually. On occasion, he crafts a matching pipe tamper.

make a better pipe. Jurkiewicz came back a couple weeks later with a completely reworked pipe and Uhle was impressed.

Carving pipes on the side, Jurkiewicz developed skills and knowledge. Uhle, to Jurkiewicz's surprise, was a skilled pipe maker in his own right who studied with Achille Savinelli in Italy. In 1978, Jack Uhle and a partner purchased Arlington Briar Company and employed Jurkiewicz part time making pipes on fraising machines. After a couple of years, Jurkiewicz returned to crafting his own pipes.

In 1994, Jurkiewicz retired to pursue pipe carving. Working out of his basement, he begins a pipe with a sketch on the Grecian or Italian ébauchon or plateau block he has oil cured. He then drills each of the holes and cuts the block with a bandsaw. He uses coarse sandpaper to shape the pipe followed by finer sandpaper. Jurkiewicz also uses a bandsaw, two lathes—one for stems and the other for pipes—several sanding discs, a sandblaster, and buffing wheels. He makes traditional pipe shapes and freehands in equal number.

Jurkiewicz crafts his stems from vulcanite and Lucite and occasionally fashions a handmade stem from vulcanite rods. Shank inserts are made from bamboo. Jurkiewicz signs his pipes "E. Andrew" and he makes about 150 pipes annually. His work is superb, with smooth and graceful lines and long and sleek stems. Retail prices for E. Andrew pipes are $125 to $800.

Alfred Baier

Located in Manchester Center, Vermont, Alfred Baier has been carving pipes for twenty years, of which the last seventeen have been in a professional capacity. Baier's carving style is very different from most carvers. Self-taught, he specializes in figural briar pipes and highly stylized freehand pipes that are unlike any

Baier's freehand pipes are some of the most unique and stylish pipes in the United States.

other pipes on the market. His figural pipes are reminiscent of the St. Claude briar pipes of the nineteenth century and meerschaum pipes of the late nineteenth and early twentieth centuries. Many of his figural pipes are carvings of heads including Abraham Lincoln and Sherlock Holmes. He also carves figurals in a story-oriented style, such as a full-length mermaid, a beaver chewing on a stick, or a squirrel leaping from a tree. Overall, his style closely emulates many of the meerschaum carving styles of the nineteenth century.

Baier uses a technique similar to openwork or filigree on some of his freehands. He carves channels and openings in the briar giving the pipe the appearance of tiny land bridges or footpaths. He uses this technique to connect the top of the bowl to the end of the shank where the stem is joined or around the front of the bowl to the foot. Whenever possible, he leaves the very attractive, original rough outer surface of the briar intact on the bowl's top and bridgework. This gives it a natural appearance and the feel of a real footpath through the woods. Alfred Baier's work is nothing short of wonderful and his talents and skills in the world of pipe carving are astonishing.

Working by himself, Baier only uses the highest quality Grecian plateau briar. Before Baier begins to carve one of the 300 or so pipes he makes annually, he completely seasons his plateau briar with a special substance he has developed to remove the impurities from the wood. He then continues with the drying process to "insure a cool, mellow and flavorful smoking experience."

Once a block is ready for carving, he roughs out the shape with the bandsaw. Baier then uses knives and files to carefully refine the shape and smooth the contours and curves of his work of art. He smoothes and polishes the wood with several grades of sandpaper. A light coat of alcohol-based stain is applied to bring out the beauty and grain of the pipe. The wood is then finished with a natural vegetable wax for a high sheen. Baier uses molded vulcanite and Lucite stems for his pipes. Alfred Baier engraves his name and date on the shank of each pipe, all of which are commissioned by clients and sold directly to them. As an option, the customer's initials can be stamped into the stem in 23-karat gold. His pipes retail in the range of $75 to $500.

Alfred Baier is a member of several regional and national sculpture and wood carving associations. In 1981, he was presented with an award for Excellence in Pipe Crafting by the La Confrerie Des Maitres Pipiers, de Saint Claude, France.

Boswell's Pipe & Tobacco

J. M. Boswell is a very busy man occupied with a family, a tobacco shop, a pipe repair shop, and a massive pipe-making operation. Boswell's Pipe & Tobacco, of Chambersburg, Pennsylvania, is a family-owned business. Boswell and his twenty-one-year-old son, Dan, carve pipes.

Boswell begins making a pipe with Grecian briar blocks. He draws a freehand design on the block and cuts out a rough out-

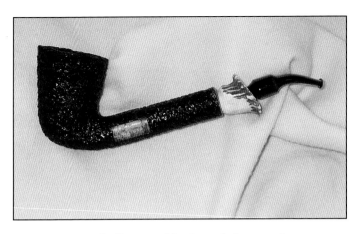

Boswell offers several finishes including smooth, sandblast, and rusticate, as well as carved designs.

line of the pipe with a bandsaw. A drill press bores the tobacco chamber and air hole, and a 50-grit sanding disc shapes the rough pipe. Boswell uses a customized "stem cutter" machine to cut the stem hole. The hard rubber and Lucite blank stems are from Italy and must be trimmed to fit the shank. He shapes the stem by heating it in hot salt in an optical box. Occasionally, Boswell uses deer antler for decoration and as inserts in the stem. Once the stem is fitted to the pipe, Boswell sands the pipe in three different phases. The pipe is stained with a brush using one of his six choices of stains and then buffed for a smooth finish. He coats the inside of the bowl with a special mixture he has developed to ease the breaking-in process.

Since Boswell's son has joined the business, annual production has increased to 6,000 to 7,000 bowls of which half are freehands and the other half are fraised. In addition to smooth finishes, Boswell also makes sandblast, rusticated, and carved designs in the pipes. His freehands are conservative with stylized bowl shapes and may be enhanced with contours cut into the sides of bowls.

Boswell can cut and sand a pipe in fifteen minutes and his son then does the finishing work. In twenty-five years of business, Boswell's Pipe & Tobacco has produced more than 150,000 pipes. Boswell's is also responsible for pipe repairs from 100 tobacco shops. His retail prices are $33.95 to $129.95 for a variety of pipe styles.

J.T. & D. Cooke

J.T. Cooke of East Fairfield, Vermont is a master carver of the briar pipe. Professionally trained as a commercial artist, Cooke made the life-altering decision twenty-five years ago to leave the work he had been so diligently trained for at the Rhode Island School of Design to hand craft pipes. In the 1980s, Cooke began to repair and restore high-grade pipes. Working on top quality pipes allowed him the opportunity to study the characteristics and quality inherit in these pipes and it gave him high expectations to match in his own work.

Working alone in his Vermont shop, J.T. Cooke produces 300 to 350 pipes per year. Quality, precision, and craftsmanship are the characteristics of Cooke's pipes. Before he begins carving a pipe, he draws the design on paper and then he studies the engineering principles of the pipe and the chemistry of the piece of briar before him. He gives an equal amount of attention and thought to the stem because the stem's design has much to do with how the pipe will smoke.

Once he has the pipe's design planned, Cooke cuts the block on a metal cutting lathe. The rest of the pipe is done by hand. He uses dial calipers to precisely measure, adjust, and check his progress against the intended classical pipe style. His hand tools include "files, sandpaper, buffing wheels, and Band-Aids." He spends on average about twelve hours on each pipe, and, as Cooke will tell you, "There are no short cuts."

Cooke uses only Moroccan briar because it is more heat resistant and similar to the old Algerian briar used in the past by Dunhill and others. It has a different feel and smell when carved as compared to other briars and produces a different finish. Cooke's sandblasting procedures are as precise and critical as his pipe making. He sandblasts each pipe three times to bring out the different characteristics of the wood. He makes his own stems from clear liquid Lucite that he hand-tints and casts into

Cooke makes one of the finest quality and attractive sandblast finishes on the market. His pipes are not only beautiful in appearance and well-balanced, but offer a superior smoke.

rods. The Lucite is turned on the lathe and then cut by hand. Cooke also smiths his own silver used in his pipes.

J.T. Cooke strives to achieve the perfect pipe. What is perfection to J.T. Cooke? Maybe one perfect pipe per year. And this perfect pipe is stamped with a small globe signifying Cooke's achievement. Cooke carves mostly classic shapes, but will occasionally fashion a freehand when commissioned to do so. His prices range from $280 to $1000.

Cristom

Based in Tampa, Florida and Italy, family-owned Cristom is the third largest pipe producer in the United States. Thomas Cristiano, head of the company, is a man who has spent most of his life around pipes. Cristom pipes are designed and partially shaped in Italy and then completed in the United States. About sixty percent of the work is done in America.

Cristiano is originally from Calabria, Italy, home to the Calabria briar. Shortly after arriving in New York at age sixteen, he began working with S.M. Frank & Company, the makers of Kaywoodie, Medico, and Yello-Bole pipes. In 1978, Cristiano left S.M. Frank & Company, and moved to Tampa, Florida. In 1980, he teamed up with his two brothers and started the new company of Calabresi Smoking Pipes. A decade later, Cristiano went into business with the Italian pipe maker Alberto Montini of Mastro de Paja, forming Mastro de Paja USA. In 1997, the part-

The hand-crafted Thomas Cristiano signature pipe is the top-of-the-line pipe from Cristom Inc.

The Cristiano is the middle line from Cristom Inc. with about 6,000 pipes produced annually.

Calabresi pipes are inexpensive with prices beginning around $12. Over 150,000 Calabresi pipes are made annually.

nership with Mastro was dissolved and Cristiano returned full time to his family business.[35]

Today, Cristiano's company makes both hand-carved and machine-made pipes. It also imports and distributes smoking accessories including humidors, lighters, and tampers. Cristom pipes are made of plateau briar from Italy, Corsica, Morocco, and Greece. Eighty-five percent of the pipes made by Cristom are classical shape; the rest are freehand. Templates are used in classical shapes to insure uniformity and purity to the style. Craftsmen will turn about 500 bowls of one template style before changing templates to produce another design. The holes are drilled and then the pipe is sanded smooth on a sanding wheel. Even though the fraising machines turn bowls the same each and every time for shape consistency, no two pipes are identical. The hand finishing and the grain of the briar is different enough to make each pipe unique. All the stems used in Cristom pipes are made from vulcanite, Perspex, or Lucite.

Overall, the quality of Cristom pipes is superb. The pipe designs, various stem treatments, and the use of metal bands lead to a refreshing variety of pipe styles. Cristiano has trained his pipe makers to maximize the finest aspects of the briar in each pipe. The flame of the grain is one of the strongest points of Cristom pipes.

Cristom makes three lines of pipes beginning with the inexpensive Calabresi, which retail from $12 to $250. This line includes filtered pipes and annual production is about 150,000 pipes. The Cristiano pipe is Cristom's middle line. Priced from

$60 to $1000, annual production of this series is about 6,000 pipes. The Thomas Cristiano Signature pipes are the finest pipes made by the Cristom Company. Carved by hand, four full-time artisans produce between 1,200 and 1,500 pipes. Prices range from $250 to $1200.

Jody Davis
Princeton Pipes

Jody Davis is new to pipe making but already his keen sense of design and skilled technique have caught the eye of several master carvers. His teacher and mentor J.T. Cooke has aided in his quick start, and both Jess Chonowitsch and Lars Ivarsson have given him guidance as well.

Based in Nashville, Tennessee, Davis has been carving for about four years now and in a professional manner for three.

*Jody Davis pipes and promotional flyer. In 1999,
Davis introduced a limited edition series of three pipes.
The pipes are named after St. Peter, St. Simeon, and St. Francis.*

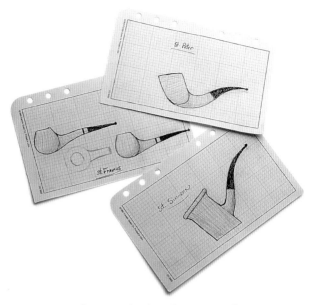

Jody Davis' sketches of new pipe styles.

His approach to pipe design blends artistic and technical methods. He takes a pipe idea and draws a sketch of his thoughts. He then fashions a pattern from the drawings and searches through his supply of high grade Corsican plateau briar for block that matches his model. Other times, when he only follows the grain of the wood, he is less technical and more fluid in his approach to pipe making.

Davis makes about forty to fifty pipes a year, all by hand. He has retooled his shop in 1999 and his arsenal includes a bandsaw, lathe, and sanding discs. He crafts a pipe in eight to twelve hours and if he is not in the shop carving, you may find him on the road in a pop rock band as a guitarist.

All of Davis' stems use the highest grade German vulcanite.

The staining and finishing of his pipes are superb and closely resemble the quality and appearance of those from Denmark. His bands and inlays are made from exotic woods and nuts and now he is searching for a legal source for ivory.

Jody Davis makes both traditional and freehand style pipes. His work is beautiful, well designed and balanced, and expertly crafted. The lines on Davis' pipes flow with grace and the curves of his bowls are delicate, pronounced, and intriguing. In 1999, Davis introduced three limited edition pipes, with ten pipes in each edition. Each pipe is named after a saint and reflects some aspect of the saint's life or teachings. Keep an eye on this budding artisan. Retail prices range from $400 to $700.

Dr. Grabow

Located in the mountainous region of northwest North Carolina in the small town of Sparta, Dr. Grabow is the largest manufacturer of pipes in the world with an annual production of about 1,000,000 pipes. Dr. Grabow is also the number one selling pipe in America.

The Dr. Grabow pipe was born in the early 1930s in Chicago when a Dr. Grabow developed an innovative screw-in-type metal filter. The metal filter provided a cooler and drier smoke. During World War II, when Mediterranean briar was not available, the Dr. Grabow workmen substituted locally harvested mountain laurel root to produce the pipes.

Dr. Grabow pipes are the only pre-smoked pipes in the world. The pipes are actually smoked by a large revolving machine that automatically loads the bowl with burley tobacco, lights it with a continuous burning gas flame, and inhales the smoke by suction. The advantages of this rare machine are that each pipe is sold with a cake lining and major flaws in the wood can be identified.

The Dr. Grabow pipe is a machine-made pipe. Using fraising machines, the workers place the briar blocks in the machines' jaws and as the blocks turn, the bowls are shaped. Other machines will shape the shanks while others form the feet and, finally, drill the holes. Workmen trim the rough edges left by the machines and then smooth the surfaces with sanders. These seasoned artisans quickly move the pipe forms in and out, twisting the wood pieces left then right as the grinders and sanders spin into the wood.

The Dr. Grabow pipe is the world's most popular pipe. Annual production numbers more than one million pipes.

Dr. Grabow pipes are all classic in shape. Several finishes are available, and the carved treatment—where the outer surface is textured to resemble tree bark—is especially popular. Each pipe is stamped "Dr. Grabow" and the stem is graced with the very familiar white ace of spades emblem. The quality of Dr. Grabow pipes is very good and on occasion a Grabow pipe is made in which the quality of the wood and the smoke will equal a Dunhill.

The Dr. Grabow people also make another line of pipes called Alpha. Ranging in price from $30 to $45, Alphas are quality pipes with a reasonable price tag. Unlike the Dr. Grabow pipe, Alpha pipes have no filters and were originally made in Israel. Made from Mediterranean briar these pipes are all classic in shape and come in five finishes including sandblast.

Fairchild Pipes

Like some other carvers, Ron Fairchild of Bellaire, Texas started out carving pipes to create a pipe he wanted to smoke but could not find. Using pre-drilled plateau straight-grain briar blocks, Fairchild created his first freehand pipe in 1978. He was elated and his interest was piqued. His early source for briar blocks and stems was Randy Wiley. Later Fairchild learned more about carving pipes from Mike Butera and Ken Barnes of James Upshall Pipes.

Using Grecian and Italian plateau briar, Fairchild draws the design of the pipe on the block and cuts out the basic shape with a bandsaw. He clamps the briar on a lathe and turns the outer surface of the bowl with a hand-held chisel. The block is repositioned and the shank is turned, so the stem hole and air passage can be drilled. Fairchild then bores the tobacco chamber to meet the air hole using bits he has personally made. Sanding discs are applied to smooth out the shape of the pipe and hand sanding provides a smooth finish. A pre-cast vulcanite or Lucite stem is fitted to the pipe and if the pipe is a high-end piece, Fairchild may add an insert or band of micarta (imitation ivory), brass, nickel, silver, or an inlay of briar or staghorn. Fairchild was one of the first Americans to use a briar inlay in the stem. All of his pipes are buffed with canuaba wax.

More than sixty percent of the pipes made by Ron Fairchild have smooth finishes. He seldom makes a sandblast or a carved finish and if he does, he leaves it unstained. Originally, he made freehands, but today, most of his pipes are of classic designs, the

Fairchild is one of only a few carvers who leaves his sandblast finish in the "raw." Your natural oils from your hands will leave the sandblast with a beautiful deep coloring.

Most of the pipes made by Fairchild are finished smooth. His best pipe is the billiard.

billiard being his best piece. His briar inlay work is well executed and very stylish. He signs his pipes "Ron Fairchild" and uses three white dots on the side of the stem. Fairchild Pipes range in price from $175 to $2,000 and only fifty to one hundred pipes are made each year.

David Jones Briar Pipes

David Jones is a multi-talented and very busy individual. His father was a carpenter and a farmer, and as Jones grew up he developed a great respect for hardwoods. Curious about applying his wood carving skills to pipes, he bought some low-grade briar blocks to experiment. Lacking in time and skills, he gave up. About twenty years later in 1986, he located high quality plateau briar in the Pipe Collector's International magazine. Using only knives, he carved about thirty pipes during the next several months. A couple of years later, he began to exhibit his work at PCI shows and the pipes began to sell. He came back to

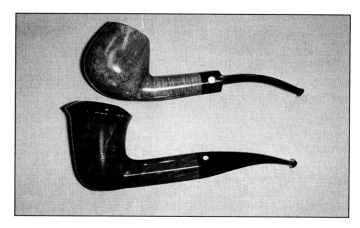

Producing about 120 pipes annually, Jones strives to bring out the beautiful grain in the Grecian and Corsican plateau briar.

Texas, bought pipe-carving equipment, and converted his garage in Texarkana into a carving shop. Sales continued to grow and in 1992, Jones built a 20-foot-by-30-foot building for his "business."[36]

Today, Jones makes about 120 pipes each year. He uses high-grade Grecian plateau briar and some Corsican and Algerian plateau briar that he has aged for five or six years. Like some of the Danish carvers, he brings his briar into the shop for an additional two-year aging process.

Jones' work reflects his very careful attention to detail in construction and design. His tools are similar to other high-quality carvers: bandsaw, lathe, drill press, knives, rasps, chisels, and a 60 grit disc for shaping. His particular strengths are his finish and stem work. Maple, bamboo, boxwood, and holly are used to provide interest and beauty to his stem inserts. Stems are made of Lucite and, when requested, vulcanite. His pipes are finished in six grades: one rustic and five smooth. The finer David Jones' pipes are well-carved, elegant looking, and offer a very dry and cool smoke. He marks his pipes with his signature and has a numbering system that dates and numbers them. Retail prices range from $195 to $475.

Kaywoodie

Based in Peekskill, New York, Kaywoodie pipes are among the oldest and most recognized pipes in America. The Kaufman brothers from Germany began the Kaufman Brothers & Bondy Company (KBB) in 1851 as a small pipe shop in the Bowery section of New York City. KBB first made the Kaywoodie pipe in February 1919. KBB also introduced another pipe called the Dinwoodie in November of that year, but discontinued it in 1924. It is believed the name Kaywoodie was created by combining the letter "K" from Kaufman and the word "wood" representing briar.

In 1930, the corporate office of KBB moved into the newly built Empire State Building in New York City. A few years later, manufacturing operations were relocated to a larger facility in West New York, New Jersey. This period marked the height of their production as the largest pipe making facility in the world. A staggering ten thousand pipes were made daily by five hundred employees. In 1938, KBB opened a successful Kaywoodie office in London in conjunction with Comoy's of London. The joint venture lasted until the 1970s, and today, Comoy's still makes several different styles of Kaywoodies.

World War II caused severe shortages of Mediterranean briar for American pipe makers. Accordingly, KBB began experimenting with briar harvested in the Santa Cruz Mountains of California called manzanita or Mission Briar. In March 1955, S.M. Frank & Co., a large pipe manufacturer, wholesaler of tobacco items, and parent company of Wm. Demuth & Co and Medico Pipes, purchased KBB and the Kaywoodie Company. During the 1970s and 1980s, S.M. Frank & Co. relocated its manufacturing facilities several times, finally settling in Tampa, Florida. Today, all of the 100,000 briar Kaywoodie pipes made annually are manufactured in Tampa and distributed worldwide.

Kaywoodie pipes are made primarily from Calabrian briar. The company uses a variety of machines—many of them more than fifty years old— in its pipe production including fraising equipment, a modified drill press for drilling mortise and air holes, tenon cutters, a barrel sand blaster, assorted buffs for scouring, buffing, and polishing sanding wheels, and a stamping arbor press. Because of the high volume of pipes manufactured by the company, stems are heated in a custom-made oven and bent by hand with jigs. Black nylon stems are used in all lines except the finer Gold Series Kaywoodies that have Lucite or vulcanite push stems.

The new Gold Series Kaywoodies, retailing from $35 to $75, were introduced in September 1997 and offer brass, briar, and Lucite stem inserts. The company also entered the mid-

The Kaywoodie Supreme is one of eight finishes available in the new "Gold Series" pipes. The pipes are identified with a gold stamped cloverleaf on the stem.

range market in 1998 with its new Handmade Kaywoodies. Using Grecian briar the pipes are made by hand to bring out the best in the briar. The pipes have Lucite push stems, are stamped with a serial number and a date and retail from $100 to $300.

Kirsten Pipe Company, Inc.

Manufactured in Seattle, Washington, the Kristen Pipe is one of the most futurist and aerodynamically designed pipes available today. The small briar or meerschaum bowl screws into the aluminum alloy stem. The ebonite or Lucite mouthpiece is connected to a long, slender aluminum tube and it slides into the aluminum stem. Kirsten is referred to by many as "the world's coolest pipe."

The origin of the Kirsten pipe goes back to 1936, when Professor F. K. Kirsten, a German immigrant from Hamburg, was told by his doctor to use a pipe instead of smoking cigarettes. The professor followed the doctor's orders and began

Kirsten offers a custom ordered pipe available in 6 bowl shapes of 3 different sizes, 3 smooth and 3 sandblast finishes and meerschaum. Stems are offered in 3 colors in full-bent, quarter-bent and straight. All Kirsten pipe parts are interchangeable and easily cleaned.

smoking a briar pipe. To dissipate the pipe's heat, Kirsten, a master machinist, developed a tube-style stem out of an aluminum alloy. His stem enlarged the inside of the tube allowing the air to cool more rapidly. He also found that his tube stem, being larger and cooler, captured more moisture, through condensation, and that the moisture removed tars, nicotine and acids that caused tongue bite.

The Kirsten Pipe may very likely be the most interchangeable pipe today. The briar is turned on a machine, and the other parts—including the stem, mouthpiece tube, tube valve, and connectors—are all precision machined. The bowl is completed by hand for a smooth, sandblast or rusticated finish and then buffed. In the event a part is broken or lost, replacement parts are available.

Today, Kirsten Pipes is owned and operated by the Kirsten family. Gene Kirsten, son of the professor, still assists in their production. Kirsten bowls are available in five shapes with straight, quarter-bent and full-bent stems. Kirsten pipes, many of them built to order, are sold very differently than other pipes. Retail prices range from $37.50 to $90.00.

Samuel Learned

Samuel Learned of York, Pennsylvania is relatively new to the field of pipe carving. But looking at his end products you would not guess that this is the case. Learned has discovered and is in love with the horn-shaped pipe. His signature lines are the "Hunter Series" and the "Millennium Horns." His main teacher and mentor has been Jim Margroum. He begins a pipe either with a design he has thought about or with a design suggested by the wood. If the pipe is a new style, he will first draw the

Learned offers traditional and freehand styling in pipes. His bamboo inserts are beautifully and skillfully created.

*The signature pipe of Learned is the "Hunter Series."
The horn-shaped pipe is offered in five styles all with antler inserts.*

design on paper, make a prototype and smoke it for evaluation. Using Grecian plateau briar or premium Corsican briar, Learned removes with a bandsaw the sides of the block and sands to inspect the quality of the wood. If the pipe shape is standard, he sketches the design on the block closely following the grain structure. He then draws the air hole and tobacco chamber on the block and drills.

Once he has installed the stem, he begins to shape the pipe with an 80-grit sanding wheel. The shape of the pipe and contours are then refined with a 120-grit wheel. Learned completes the sanding with a 600-grit paper. He then polishes the pipe with a triple E compound and sets it aside for a couple of days. After staining, the pipe is once again given a rest, this time for a month before it is polished with a natural wax. Sandblasting and rusticating are applied after the shape has been refined.

Samuel Learned's carving style is very Danish and most of his pipes have short, rounded, block-style bowls. He is fond of producing forward slanting bowls on many of his pipes, the horn pipes in particular. His lines are graceful and the curves are gentle and very rounded. He makes excellent use of bamboo and antler inserts. Learned pipes feel very comfortable in the hand. He produces around 300 pipes annually. Retail prices range from $100 to $1,200.

Lucille Ledone

"Custom carved pipes for discriminating smokers" is the motto of Appleton, Wisconsin's Lucille Ledone, Cille to her friends and customers. Women are not new to the world of pipes. However, Ledone holds a very special position in pipedom. Not only does she approach the pipe from a female perspective, but many of her customers are female. She understands very well their expectations and comfort levels with pipes. Most of the pipes she makes for women have smaller bowls but longer stems.

An attorney by training, Ledone began carving pipes after reading Richard Hacker's *The Ultimate Pipe Book*. With the help and guidance of Alfred Baier of Manchester Center, Vermont, she began to pursue pipe making professionally. After two to three years of carving, Ledone began to really understand the mechanics and art of pipe making.

Ledone uses only the highest quality plateau briar from Greece. To ensure a "cool, clean smoke," she seasons the sevety-five-year-old briar in a special substance that removes impurities in the wood. She doesn't really begin with a design in mind, she just lets the beautiful flame patterns guide her as she shapes the pipe. All of her carving is done by hand using only Danish and

Swedish wood working tools, a Dastra knife, her favorite, smoothes the wood with hand sanding for about eight hours. To complete the pipe, she uses a polishing wheel for buffing the finish.

Ledone creates freehand-style pipes. Some of her pipes present bold, blocky bowls with projecting angles while other pipes exhibit more of a fluid motion with gentle curves and graceful angles. She tends to flare the shank area of the pipe as it meets the stem, a nice design feature. She is fond of using long stems, up to eight inches on her churchwarden-style pipes. The influence of Al Baier can be seen in her work as she forms the bridgework of flowing briar from bowl to shank.

Because all of her work is done by hand, Ledone spends eighteen to twenty-four hours on each pipe, and signs every pipe with a hand-engraved "Cille" followed by the year. Her annual production is only forty-five to sixty pipes and her retail prices are $95 to $350.

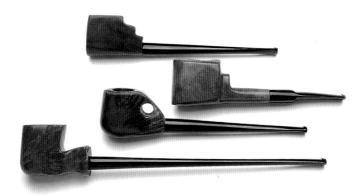

Ledone is partial to long stemmed pipes. Each of her 45 to 60 pipes, which she hand-crafts in a year require 18 to 24 hours to complete.

Andrew Marks

Andrew Marks had been a pipe smoker for many years before he decided to pursue pipe making. He became intrigued with the idea of pipe making when the Nørding pipes first came to the United States in 1969. Marks was very familiar with the English, French, and Italian pipes, but it was the Danish designs that spurred him in the direction of carving.

While living in California, Marks met Mark Zeavin, a pipe sculptor, and his approach to pipe making became more artful. His first tools were a vise, files and rasps, and an electric drill. After returning to Cornwall, Vermont, he later visited several master carvers in Denmark—including Anne Julie, Poul Hansen, Sixten Ivarsson, and Former Nielsen—all of whom influenced Marks' carving style.

Marks begins by drawing a simple design on an extra quality block of plateau briar from Greece. Unlike most carvers, he then drills the tobacco chamber and air passage to ensure the briar has no major hidden flaws. Using a bandsaw, he removes the excess portions of the block and shapes the pipe with a sanding wheel loaded with 36-grit sandpaper. As he continues to shape and refine the pipe, he changes the sanding paper to 320 grit and then 400 grit. If he desires a "super smooth finish," he will use 600-grit paper. The pipe is then wiped smooth with a crocus cloth. Marks then hand-rubs the pipe with extra virgin olive oil and on some occasions he will finish the pipe with a wax.

Marks' designs and presentation of the briar grain are attributes from his earlier contacts with Danish master carvers.

Like the Danish masters, he often designs and shapes the pipe according to the grain of the briar. Quite often he chooses to work without a design, allowing the grain to guide him. He may well turn the block on its side and work it for its exquisite birdseye grain. His finishes are also dictated by how the grain will respond to the different materials. He hand-shapes his stems, buffs a mirror finish, and bevels the air hole where the tongue touches the stem opening. His bits are wide with a thin profile for maximum comfort. Marks spends from a half-a-day to three days on one of his creations.

In the 1970s and early 1980s, Marks made beautiful and graceful freehands. He was fond of leaving the rough outer surface of the briarroot exposed as a prominent design element of the pipe. Today, he prefers the classic shapes in pipes. Equally beautiful, his classic designs are intriguing, clean and well proportioned. He strives to bring the straight grain to prominence and give the pipe balance and elegance.

Today, Andrew Marks makes less than fifty pipes a year, all custom ordered. In the past, when he was a full-time carver, he made between three and four hundred pipes. In all he has crafted almost fifteen thousand pipes. His pipes retail for $350.

Mr. Groum Pipes

One of the more unusual pipe carvers is James Margroum. This self-taught carver demonstrates the lighter side of pipe making, he makes fun pipes. These smokeable creations are made of plateau briar, and are carved in the form of animals and objects.

The apple pipe allows Margroum to bring a little humor into pipe making. Each is complete with a mouth-size bite and part of a worm!

Margroum tries to bring laughter and a great sense of humor into the very serious pipe world.

Based in Hanover, Pennsylvania, Margroum makes about two hundred pipes a year from Grecian plateau briar. He cuts off a piece of briar and finds the grain of the wood. At this stage he formulates a design for the wood and sketches it onto the block. He uses a bandsaw to rough cut the pipe and then drills holes for the tobacco chamber and stem. He then employs hand-held motor tools, burrs, and dental drills to shape and carve the briar. Margroum has about six drills each loaded with a different bit to save time. Using sanding discs, he spends a great deal of his time smoothing and polishing the pipe. When asked how he carves a pipe, Margroum jokingly answers "I cut away everything that I don't want, and what is left is a pipe."

Margroum will cut or sand six to twelve pipes at a time and may work on a pipe for more than a year. Mr. Groum's pipes are "shapes that suggest objects used or seen in everyday life." Even the most figural pipe is a smokeable object. His ideas can be quite humorous. He's made a commode pipe for a plumber, a miniature Mack truck for a truck driver, and golf ball and baseball pipes for sports fans. His trademark pipe is an apple with a bite taken out of it and half of a worm! The stem and a leaf form a windcap.

Not all of Margroum's work is fun and games, however. He makes very serious freehand pipes that bring out the beauty and quality of the briar. Some of the freehands have flared bowls and shanks. In others the bowl extends as a bridge over to the stem, forming a beautifully contoured opening for the thumb. He uses stems from vulcanite and Lucite and occasionally uses deer antler for shank extensions.

James Margroum pipes are available through his website. The prices range from $40 to $1000.

Clarence Mickles

Clarence Mickles has always had an interest in expensive pipes but could not afford to purchase them. So, in 1976, he began making his own pipes as a hobby, and in 1995, he turned his hobby into a profession. Most of the pipes he carves in Park Forest, Illinois are freehands.

Mickles draws a design he has been thinking about on a briar block and cuts it out with a bandsaw. Customers can also send in their own designs or can explain how they want their pipes to look and Mickles will follow the designs or requests. He applies the sander to work around the grain and shapes the pipe.

Mickles makes a large pipe. Most of his work is traditional in design or very conservative freehands.

Once the pipe has been shaped and drilled, he boils it in water until the water runs clear. This sap removal process insures a finer smoke. His final sanding is with 400-grit sandpaper. If he is creating a pipe of exceptional wood, he uses 1500-grit sandpaper to bring out the quality and finish of the briar. He uses vulcanite and Lucite rods and Lucite blanks (pre-shaped stems) for his pipe stems.

Mickles prefers the freehands, but he also makes billiards, pokers, rhodesians, and other classical styles. Some of his freehands are based on traditional shapes, but the bowls or stems are crafted with a hint of a contemporary flavor. In his freehands pipes, Mickles sometimes slants the bowls forward or softens its edges with subtle swirls or flute cuts, giving the pipe a distinctive but soft, masculine look. His signature mark is two white dots located on the center and side of the stem. Mickles also does beautiful sandblast and rustication finishes and uses an occasional engraved gold band.

Mickles produces about 300 well crafted, good smoking pipes annually. His retail prices are $150 to $350.

Elliott Nachwalter Pipestudio

Nestled between the Taconic and Green mountains of Vermont and overlooking the Battenkill is the Arlington pipe artisan studio of Elliott Nachwalter. Nachwalter has had an interest in tobacco and pipes for many years. As a senior in college, he

Nachwalter's pipes have conservative freehand designs or are conversative intrepretations of the classic shapes. In either case, his pipes have a strong Danish appearance.

opened a pipe store in Vermont. In 1972, he began to carve pipes. Having studied art history and photography and taught photography for a period, he realized his interest and creative talents could best be served in pipe making. Nachwalter has studied with Andrew Marks and even shared a pipe store with him. Nachwalter also worked with Finn Meyan Andersen, a pipe designer and maker for W.Ø. Larsen in Denmark.

Nachwalter uses the very finest Ligurian and Calabrian plateau briar from Italy and Corsican plateau briar and air cures it for up to six years. He begins a pipe by drawing its shape on paper. He then searches through his briar supply for the one piece with the appropriate grain that will allow him to create his vision. He applies the sketch to the block and cuts it with a bandsaw. A metal cutting lathe that has been converted for pipe making drills the holes. Nachwalter uses a sander from Denmark to shape his pipe and does the final sanding by hand. Nachwalter makes either a hand-cut stem from acrylic, cumberland, or black vulcanite on request, or he will use an Italian-made stem he has designed. If he chooses a natural finish, he rubs the pipe with olive oil to complete it.

The pipes from the hand of Elliott Nachwalter are contemporary interpretations of classic pipe shapes. His work is conservative and some of it exhibits an expected Danish influence. His pipes have a wonderful symmetry and balance about them. The pipe shapes and lines are very graceful, flowing, and appealing. He is skilled in bringing out the beauty of the grain and his finishes are superb. Retail prices begin at $250 and top at $3,000.

Denny Souers

Denny Souers of Columbus, Ohio is a veteran pipe artisan of thirty years whose pipes exhibit a wonderful artistic aura. He produces mostly freehands, truly interpreting the word "free" in his style of pipes. His designs are fluid, and his finesse with the wood results in graceful lines and contours as the wood flows from bowl to stem.

Before Souers began carving pipes, he attended college for a couple of years and worked as a draftsman, followed by employment at DuPont in the plastics department and finally Owens

Corning. His interest in pipe smoking stemmed from his cigarette habit, but he could not afford good quality pipes. Realizing his love for tobacco and the cost of a good pipe, his wife purchased him a briar kit. His first pipes went to friends as gifts. He soon showed his pipes at an arts and crafts fair and sold several of them.

Souers' eye for design is most evident in his pipes. Particularly noteworthy is his masterful treatment of bridges and his effective utilization of the natural burl surface. A variation to the bridge is what he calls the web, the section between the bowl and the top of the shank. Souers leaves this area solid, but first shapes contours and folds in the web. The Sidewinder pipe, for example, has the stunning presentation of natural burl along its sides.

Having carved pipes for thirty years, Souers has experimented with many tools and techniques. He begins with a high

Pipes made by Souers are one-of-a-kind freehands. His form and lines are exciting, unpredictable, and offer a sense of great satisfaction.

grade of Grecian briar. After curing it for at least six months in his shop, he begins the six-hour ordeal of pipe making by drilling the tobacco hole followed by the air hole. He trims the block with a bandsaw, removing slabs of wood. His main cutting tool is a one-quarter horsepower motor armed with heavy 3 inch and 4inch diameter cutting wheels. He quickly cuts the wood forming his shape with 10 to 25-grit carbonite wheels. Once the shape has been roughed out, he switches to a regular sanding disc that has medium density foam underneath enabling him to work the contours and folds of the pipe. The final shape of the pipe becomes apparent as he switches between 220 and 320-grit sandpaper.

Souers has discovered a special process he applies to his pipes during the sanding phase that gives the grain a smoother finish and allows the wood to breathe easier. His wife, Rusty, hand sands all of his pipes, applies stain, and polishes the pipe with canuaba wax on buffing wheels. She also matches the hard rubber or an occasional Lucite stem to the pipe. Rusty performs the final inspection of all Denny Souers pipes.

Most of Souers' pipes weigh less than two ounces. His use of contours and shapes gives the pipes a larger appearance. The pipes he makes with bridges and webs are custom designed for left-handed or right-handed smokers

Now that he and Rusty have retired, plans are to increase pipe production from the current level of 150 to 200 pipes annually. They plan to travel more and attend more pipe shows. Retail prices range from $140 to $500.

Trever Talbert

Trever Talbert is new to pipe carving and brings a fresh approach to the process. A freelance illustrator and writer based in Thomasville, North Carolina, Trever carved his first pipe about six years ago as a whim, as a way to relax and be creative.

In a very short period of time, Talbert's pipes took on a professional appearance and his friends and acquaintants wanted to purchase them. Talbert, a collector of fine pipes, initially would not consider selling his carvings because he thought the quality of his pipes and the time needed to produce a fine pipe would be out of line with people's expectations. When he won a pipe making contest hosted by *Pipes & Tobacco* magazine, however, he decided it was time to carve pipes for sale. Reflecting on that decision, Talbert comments "I was confident they were equal to the high-end pipes I enjoy and I liked making them."

Talbert makes his pipes entirely by hand using only a disc sander, drill press, two belt sanders, a metal lathe for stem work, hand files, and a Dremel tool. He purchases high quality briar from Greece, Italy, and Sardinia. Talbert has an interesting way of storing his well-aged briar: he stores it in a window facing south to "bake" it for two to three years before carving.

Trever sketches "for days and days" perfecting his designs for "styles that are interesting visually yet still maintain solid engineering." He is extremely careful with the designs and how they interact with the airflow and temperature of the smoke. He begins with a piece of briar that inspires him, then draws the

Talbert makes about forty pipes a year, with both smooth and sandblast finishes.

design on the raw block, and slowly removes the wood. He spends hours sanding the pipe by hand and applies stain in layers to bring out the character of the wood. After the natural waxes are applied, he either hand-cuts the stem from German or Dutch vulcanite rods for the high-grade pipes or uses acrylic stems for the low and medium grade pipes. Talbert spends six to forty hours per pipe, depending on the complexity of the design.

No two of Talbert's pipes are alike, and he prioritizes function over style. Still, Trever Talbert handcrafts a beautiful pipe. He uses sweeping lines and graceful curves. He takes traditional shapes and adds his own subtle interpretation of the briar grain. He maximizes the grain impact of each block giving the pipe a striking pose. His gentle use of design features go a long way in separating his talents and eye from other carvers. Since he is very new to the world of pipe making, his talent and abilities have only been tapped. His annual production as a part-time pipe artisan is thirty to forty pipes. Presently, all of his pipes are sold by telephone or through his website. His prices range from $180 to $200 for rusticated pipes, $400 for smooth finishes, and $600 to $1,000 for ultra-high grades.

Mark Tinsky
American Smoking Pipe Company

Mark Tinsky and his friend, Curt Rollar, began their careers in pipe making by cleaning briar for Tinsky's neighbor, Jack Weinberger. After Tinsky finished college, he joined Rollar who had already become a full-time pipe maker. Tinsky and Rollar establish their own pipe company, American Smoking Pipe Company in 1978.

The early days were lean, and they had very little capital and absolutely no briar stock. They worked for another pipe carver taking briar in lieu of money and lived on a farm with a makeshift workroom. Their first bandsaw and drill press were purchased from Sears. Later, Tinsky and Rollar moved the business to eastern Pennsylvania. After Rollar left to pursue another career, Tinsky continued the business on his own, relocating his shop to Pocono Lake, Pennsylvania and adding pipe repair to the agenda.

Today, Tinsky produces about 500 pipes annually. His wife, Maryann, repairs pipes for about one hundred tobacco shops. He has an additional two assistants who polish and finish pipes. He uses high-quality Grecian plateau briar and custom-made black Lucite stems from Italy. Tinsky begins a pipe by cutting out the shape on a bandsaw. Using a lathe, he then drills the holes and sometimes refines the bowl shape with a lathe, hand chisels, or a sanding disc. His stems are also cut on the lathe and the detail shaping is done with files.

*Tinsky makes many different styles of pipes.
Most of his pipes are traditional in styling and his freehands
have conservative overtones.*

Most of the pipes Mark Tinsky makes are classic in shape and are well proportioned and balanced. He makes only a few freehands and these are very Danish in appearance. Each year, American Smoking Pipe Co. offers a limited edition Christmas pipe. Tinsky identifies each pipe with a stamped silver star surrounded by a briar ring. He also sells raw briar blocks to customers for carving. Retail prices begin at $100 and top at about $500.

Von Erck's Pipes & Repair

Lee E. Erck is a different kind of pipe carver. Working out of his small shop in the northwoods of Michigan, Erck captures the raw beauty of the great outdoors in his rugged freehand pipes.

His use of strong lines, bold angles, and deep contours form a pipe for serious smoking. His tendency is to form slender stems and tall bowls. He favors smooth finishes but some of the contouring leaves the impression of a carved finish. On some pipes, he applies rustication in very select locations giving his pipes a two-toned effect.

Erck has been carving since the mid-1980s. His interest in carving grew out of pipe repairs. Having studied the mechanics of pipe craftsmanship from repair work, Erck took it upon himself to learn the art of pipe carving. He begins with a block of Grecian plateau briar or occasionally Italian or Algerian briar and handcrafts one pipe at a time. Each pipe is unique and created with only a bandsaw, belt sander, drill press, and buffing wheels. Erck says his designs come from the wood itself, and stresses that he wants to create pipes with "eye-catching" appeal.

Erck's signature is the random use of rustication. Combined with the gentle curves of the briar, the pipe takes on the appearance of staghorn.

Before Erck marries a stem to the pipe, he oil-cures the wood and then dries it to insure a pleasant smoke. He hand crafts the stem for the pipe using Lucite and vulcanite from Germany. The final stage in making a Von Erck Classic is stamping his name and a serial number that includes the date and the pipe number. Lee Erck makes about four hundred pipes annually and his prices range from about $100 to $750.

Roy Roger Webb

In the Smoky Mountains of east Tennessee lives an unusual pipe carver. Roy Roger Webb carves a different kind of briar pipe. Webb carves in minute detail the faces of spirits on the bowls of pipes. Many of the spirits are characters out of mythology and others are characters he has developed himself. For example, the Four Winds pipe is shaped like a square with a different face, or wind, on each side.

Intrigued with carving because of his father's influence, Webb purchased a set of wood working tools and began carving about twenty-six years ago. Webb began to wonder if his spirit faces would sell as pipes. The spirit pipes did sell and he continued to purchase pre-drilled briar blocks.

As Webb developed his skills as a pipe carver, he considered the purchase of an expensive drilling machine to bore holes in the pipes. Webb decided against the machine, however, and to this day he drills his own holes by sight. Webb uses pre-cured

Roy Roger Webb's spirit faces represents some of the finest figural carving in the pipe world. His carvings take on the appearance of life-like characters he has created.

Grecian plateau briar that is one to two hundred years old. His premium briar costs him between $50 and $100 a block. He begins his work by determining the best angle to drill the bowl and air passage. After the holes are drilled, he carefully carves the spirit face on the pipe. Because his work is so detailed, he uses a Dremel tool to work the main features and finishes the finer facial and hair details with small knives. The eyes are the most difficult features to create. He uses a magnifying glass as he carefully works the wood around the eyes.

Other tools used in his carving arsenal include U- and V-shaped gouges. Webb finishes the pipe with sanding papers. He begins sanding by hand the knife marks with 100-grit paper, followed by 180-grit and then 220 grit-paper. He uses no stain on his pipes; instead, he finishes the wood with a wax and fits the pipe with a pre-formed acrylic stem.

Webb has been carving pipes for about thirteen years. During those years he has been both a part-time and full-time carver. As a full-time pipe artisan, he carved about 150 pipes annually. Today, he produces about fifty pipes each year. Because all of his work is by hand and the spirit faces involve tedious detail work, each pipe requires about twelve hours of work.

His retail prices range from $150 to $300. Recently, he has begun to group his spirit pipes in packages. A set of eight to ten pieces includes spirit pipes like Father of Time, Merlin, Faith in Time, Sleeper, Dreamer, and Winker, and sell for $1200 to $1500 a set. Webb numbers each pipe he makes and includes a certificate of authenticity.

Steve Weiner

Based in Fox River Grove, Illinois, Steve Weiner truly crafts a beautiful and balanced classical pipe. He began carving pipes after inheriting his grandfather's woodworking tools and machinery in 1990. He initially experimented with pre-drilled kits and then moved up to plateau block briar. However, his abilities were not polished and Weiner practiced pipe carving in his spare time until January 1997 when he transformed his income-producing hobby into a full-time career. By many pipe masters' clocks, Weiner is still in his adolescence stage, but by most masters' product standards, his quality and style are mature.

Steve Weiner prefers plateau briar from Italy for his pipes. He also uses briar from Greece and Corsica. He begins his work

Weiner creates handsome rusticated pipes.
He also makes pipes with natural and sandblast finishes.

with tracing an outline on the block of wood from a pre-designed template and then rough-cuts the shape with a bandsaw. He draws lines on the block pinpointing the bowl interior, air hole, and mortise joint for the stem. Weiner uses a drill press to bore out the air hole and the bowl interior. The lathe cuts the mortise joint and finishes the upper two-thirds of the bowl's exterior.

Weiner uses stems made from black Italian Lucite. He then inserts the stem into the shank's mortise and finishes shaping the pipe and stem with sanding discs and sandpaper. It is at this point in the process that Weiner decides on whether the pipe's finish will be smooth, carved, or sandblasted. The final step in the process is sanding, staining, waxing, and buffing. On his smooth pipes, he uses five different sanding papers beginning with 150-grit sand paper and ending with 1000-grit. Finally, this pipe carver goes one step further in the final presentation of his product. In addition to each pipe being sleeved and boxed, Steve Weiner includes a card with a date of manufacture, a pipe identification number, and his own personal signature. Prices range from $180 to $475.

Tim West

Tim West of Columbus, Ohio began carving pipes in 1967 at age seventeen, after purchasing some old briar blocks and mouthpieces from his local tobacconist. West stopped carving shortly

Above: The "Trophy Series Super Ball" is one of the most difficult pipes West crafts. The pipe bowl is perfectly round and offers West many challenges.

Below: West makes a variety of pipes. Many of his designs are based on classic shapes and the nuances of the briar allow him to be creative with the final design.

afterward, but returned to pipe carving in 1975, this time as a full-time carver under the name of T. M. West Pipes Ltd. His tiny ten foot-by-ten foot shop was in a craftsman bazaar. His production was about forty custom-made pipes per month with a price tag of $50 to $75 each.[37] In 1980, West added a tobacco shop—Tim West Pipes and Tobacco—to his business. After eleven years, he sold the tobacco shop, but continued producing hand-

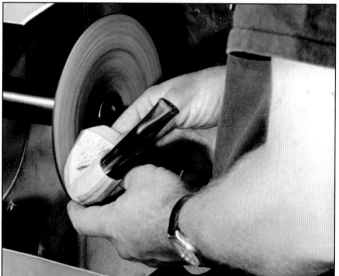

West cuts out the shape of a pipe with a bandsaw and smoothes the briarwood with a sanding disc.

carved pipes. Today, he works out of the basement in his home and produces a staggering one hundred or more pipes per month. West works basically by himself, although his stepfather, Jim Seman, comes in about three hours a day and serves basically as an apprentice and his sixteen-year-old son Collen helps with sanding and polishing pipes.

When West began carving pipes, he used simple woodworking tools. He drilled out the tobacco chamber, air hole, and mortise joint all by hand without the aid of a drill press, and did all of his sanding by hand. Today, he has modernized somewhat and this has allowed him to increase production. He now has a bandsaw for cutting, a boring machine for drilling and power sanding discs for shaping and finishing the pipes. West still guides each machine with his hands as he cuts, bores, trims, and sands the pipe.

Tim West prefers to be known as a pipe "shaper" and not a carver. He argues that carvers actually carve figures of people or animals in pipes with hand-carving tools, while shapers cut, trim, and finish wood with power lathes, turning machines, and sanders. One thing West does as an independent shaper that sets him apart is his technique of working on pipes in stages. He will perform a task on a group of about fifty pipes before he moves to next stage of production.[38]

His pipes are beautiful and their styles—a cross between classical and freehand—are, for the most part, very straight forward and practical. Tim West pipes have clean lines and most have smooth, natural finishes. Some pipes have needle point finishes. For these pipes, West uses a special tool with tiny rods of pointed steel to gorge out tiny bits of briar producing a rusticated effect. His simplest pipes take only thirty minutes to shape and finish; the majority require one to three hours. Tim's low-end prices are $35 to $100 and custom and high-end pipes command $100 to $850.

Randy Wiley

Randy Wiley has been carving briar pipes for almost twenty-five years. Based in Tampa, Florida, he was formerly a carpenter and had pursued an art degree in college. While working on a renovation project at the Hillsborough County Museum in Tampa, he met, "Pappy" Ray Holt, a wood sculptor, Wiley teamed up with Holt and served as his apprentice for two years. He was trained in the use of specialized woodcarving tools and developed a keen knowledge of woodcarving.

After the apprenticeship ended, Wiley joined a local pipe making company where he worked for two years before branching out on his own. He has been carving pipes ever since. His annual production is about one thousand pipes made of Grecian plateau or ébauchon briar. The briar dictates much of his pipe design. Wiley prefers a natural smooth finish and when he encounters problems in the wood, he redesigns the pipe to remove the flaw. As he sands, he applies stain on the wood to bring out the grain, then he sands again. He will not use putty for fills nor will he sandblast a pipe. Wiley believes that carvers must be fluid and willing to change as they carve pipes.

Many of his pipes have some rustication for which he uses tools he has specifically designed to pick and gouge the wood to form a tiny, pebble-like texture. He uses hot oil to bring out the natural beauty in the wood. His high-end pipe, the Ovation, is hand-sanded to an ultra smooth finish. All of Wiley's pipes are stained with custom mixed stains and buffed with canuaba wax.

Wiley makes pipes in the mid- to high-end range. The high-end pieces take longer to craft and the quality and attention to detail is reflected in their prices. His current work is classical in nature. In his earlier days, Wiley produced more freehands with intricate shapes and less traditional designs. Overall, he makes a beautiful pipe by combining classical elements with individual interpretation and design. Randy Wiley's retail prices for standard pipes are $85 to $500 and his Ovation line is $600 to $1,000.

Of the three main lines Wiley produces, the Ovation is his top-of-the-line pipe. Exceptional in grain, design, and craftsmanship, the Ovation series offers beautiful freehands with outstanding grain patterns.

Glossary

AMBER: Amber is petrified pine resin formed millions of years ago. Amber was an important material to the pipe world in the 1800s as a source for mouthpieces, stems and occasionally pipe bowls.

APPLE: A classic pipe bowl shape resembling the roundness of an apple.

BANDSAW: A modern day saw with a long, continuous-loop blade that resembles a band or ribbon. Bandsaws are used in pipe making to rough out the shape of a pipe in briar.

BILLIARD: Billiard is one of the most common pipe styles. A Billiard pipe has a round and tall bowl even in proportion with a straight shank.

BRIAR: The wood from the burl or protuberance found at the base of the heath tree and used in pipe making. Briarwood is found in the Mediterranean Sea area, and the major briarwood producing countries include Corsica, Italy, Greece, Morocco, Algeria, and Spain.

CANADIAN: A classic pipe shape with a long shank, pencil or oval shape, with a short stem and billiard bowl.

CANUABA WAX: The basic ingredient in most waxes using sap from the canuaba tree of Africa. The wax applied on pipes enable the pores to breathe and protects the wood from dirt and stains.

CHIBOUK: Popular in the nineteenth century, the Chibouk was a Turkish pipe that had a long stem, terra-cotta bowl, and a rounded amber mouthpiece. The more expensive Chibouks were decorated with silk and gold.

CLAY PIPE: Clay pipes were first made by the Native Americans and consisted of clay molded into a pipe and fire-hardened. The Europeans, beginning with the English and the Dutch, used kaolin for their white clay pipes. The peak of clay pipe production was in the eighteenth century.

CUMBERLAND STEM: A stem made of composite materials giving it a vein or striped appearance.

DUBLIN: A classic pipe shape with the bowl tilting slightly forward.

ÉBAUCHON: The center portion of the briar burl characterized as having an undefined grain pattern. Ébauchon blocks are used primarily in fraising machines for less expensive pipes.

EBONIZED BONE: Ebonized bone is ox bone colored or dyed black to resemble the rare and expensive wood, ebony. In the nineteenth century, many pipes used ebonized bone in pipe stems, reservoirs, and mouthpieces.

FIREWOOD: Term referring to carved pipe blanks that are thrown away or burned because of a major defect in the wood.

FRAISING MACHINE: A pipe-making machine that shapes the bowl, shank or heel of the pipe by turning the block of wood while a chisel or knife cuts the wood automatically. The most automated fraising machines have templates to guide the cuts.

GLASS PIPE: Glass pipes were in vogue from about 1790 to 1820 in England and Italy. The major manufacturing centers were Bristol and Nailsea, England and Venice, Italy. It is uncertain if any of the glass pipes were smoked. The pipes were largely fashioned as trade signs for glassmakers and tobacconists.

HORN: The horn of oxen has been a popular pipe material. In the nineteenth century, horn sections were used for making pipe stems, mouthpieces, and joining parts of the stem to the pipe. Today, many carvers prefer staghorn as shank and stem inserts.

HORN-SHAPED PIPE: A pipe designed in the shape of a horn.

LATHE: A machine that turns the wood or metal in a fast speed while one cuts or carves the material with a stationary blade. Some pipe carvers use the lathe with a drill bit to bore the pipe bowl and the mortise in the shank.

LOVAT: A classic pipe shape with a long shank like a Canadian but with a saddlebit stem.

MANZANITA: A small scrub or tree of the heath family found in the western United States. During World War II, the German occupation in the Mediterranean Sea area ended briar exportation and manzanita was used as a substitute.

MOUNTAIN LAUREL: An evergreen scrub native to eastern North America. This member of the heath family was used in the nineteenth century as a source of pipe material in America and saw limited use during World War II.

MEERSCHAUM: A clay-like, whitish mineral used in making pipe bowls. This very soft material is nearly heat resistant. Meerschaum used in pipe making is mined in an area about twenty miles east of Eskisehir, Turkey. Meerschaum is a German word meaning "sea foam."

MORTISE: An opening in the shank of the pipe where the stem fits snugly into the bowl.

MOUTHPIECE: The part of the pipe the smoker places in his mouth to smoke and balance the pipe. The mouthpiece attaches to the shank or the stem of the pipe. In today's nomenclature, the term stem has almost become totally interchangeable with the word mouthpiece.

NEEDLEPOINT WORK: Specialized tool used to gouge and pick out tiny bits of briar leaving a rusticated appearance. The tool is made of ten or twelve $\frac{1}{16}$ inch steel rods sharpened to a point and banded together in a cylinder with a T-shaped handle. The center holds four rods surrounded by six or eight rods. Another version of the tool has five tiny steel rods in a circle, but the center is left open.

JEAN NICOT: Nicot was the French Ambassador to Lisbon, Portugal, and introduced tobacco seed and snuff to the French Queen Mother, Catherine de Medici, in 1560. The scientific name of tobacco, Nicotiana, is named in his honor.

NICOTIANA RUSTICA: A plant of the nightshade family from the New World. Rustica was the original tobacco used in North America by the Native Americans. It has a strong, bitter taste and the Spanish referred to it as "like smoking hot peppers." It has small yellow to dark green leaves that are oval to round in shape.

NICOTIANA TABACUM: The scientific name for tobacco. Tabacum has large, lanceolate to oval leaves located close together on a stalk. The plant is originally from South America and is known for its mild taste. Today, it is the most important commercial grade tobacco grown.

PLATEAU BRIAR: The burl of the heath tree provides two cuts of wood. The outer cut is known as plateau briar and is distinguished by tight, straight grain.

PORCELAIN PIPE: In the eighteenth and nineteenth centuries, pipes with porcelain or China bowls and reservoirs attached to long wooden, horn or bone stems were popular in Europe.

JOHN ROLFE: Rolfe introduced the milder tobacco, Nicotiana tabacum, to the Jamestown settlement in Virginia between 1610 and 1612. The smuggled seed from Trinidad gave the English colonists their first cash crop and a foothold in the New World.

RUSTICATION: A surface treatment of a wood pipe to give it a rough texture. The wood is picked, chiseled, or gouged to form a texture similar to an orange peel.

SANDBLAST: Many briar pipes are sprayed with sand leaving a textured surface. Sandblasting leaves the appearance of the grain intact but with a pitted or pebble texture. Sandblasting removes some of the weight of a pipe and technically a sandblasted pipe should smoke cooler.

Sir Walter Raleigh: Raleigh popularized tobacco smoking in England. He was a favorite of Queen Elizabeth I and was a military adventurer, poet, and a courtier. Because of his loyalty to tobacco and the Queen, he fell out of favor with King James and was eventually executed.

Shank: The area of the pipe that connects the bowl to the stem or the mouthpiece.

Stem: In antique pipes the pipe stem connected the shank to the mouthpiece. Today, many people use the word stem for the term mouthpiece.

Tenon: The projecting part of the stem that fits snugly into the mortise of the shank of the pipe.

Windcap: A hood, usually of metal, or cover over the top of the pipe bowl. The windcap prevented ashes and sparks from blowing out of the pipe.

Water pipe: In Africa, the Middle East, and Asia, pipes with a water chamber are popular. The water, or in some cases wine, acts as a filter to remove the harshness of the tobacco and provides a cooler smoke.

Suggested Readings

Books

Armero, Carlos. *Antique Pipes (A Journey Around a World)*. Madrid, Spain: Tabapress, 1989.

Ayto, Eric G. Clay *Tobacco Pipes*. Shire Album 37. Buckinghamshire, England: Shire Publications Ltd., 1979.

Balfour, Michael. *Alfred Dunhill: One Hundred Years and More*. London: Weidenfeld & Nicolson, 1992.

Cole, J.W. *The GBD St. Claude Story*. London: Cadogan Investments Ltd., 1976.

Doll, Emil. *The Art and Craft of Smoking Pipes. A Brief Manual and Reference Work*. New York: H. Behlen & Bro., Inc., 1947.

Dunhill, Alfred. *The Pipe Book*. London, England: A. & C. Black, 1924. Revised edition. New York: The Macmillan Company, 1969.

Dunhill, Mary. *Our Family Business*. London: The Bodley Head, 1979.

Ehwa, Carl, Jr. *The Book of Pipes & Tobacco*. New York: Ridge Press, Random House, 1974.

Fisher, Robert Lewis. *The Odyssey of Tobacco*. Litchfield, Connecticut: The Prospect Press, 1939.

Fresco-Corbu, Roger. *European Pipes*. Surrey, England: Lutterworth Press, 1982.

Goes, Benedit. *The Intriguing Design of Tobacco Pipes*. Translated by Lysbeth Croiset van Uchelen-Brouwer. Leiden, The Netherlands: Pijpenkabinet, 1994.

Gregorio, Joseph M. *How to Make Smoking Pipes*. New York: Albert Constantine and Son, Inc., 1971.

Hacker, Richard Carlton. *The Ultimate Pipe Book*. Beverly Hills, California: Autumngold Publishing, 1984.

Helme, D. *The Clay Tobacco Pipe: An Illustrated Guide*. Durham, England: Brian J. Hewitson (Printing), 1978.

Hymann, Robert K. *Tobacco and Americans*. New York: McGraw-Hill, 1960.

Jeffers, H. Paul. *The Perfect Pipe. A Celebration of the Gentle Art of Pipe Smoking*. Short Hills, New Jersey: Buford Books, 1998.

Kiernan, V.G. *Tobacco: A History*. London: Hutchinson Radius, 1991.

Levárdy, Ference. *Our Pipe-Smoking Forebears (Pipázó Eleink)*. Translated by Andrew Rouse and Imre Eliás. Budapest, Hungary: Pécs-Velburg, 1994.

Liebaert, Alexis and Alain Maya. *The Illustrated History of the Pipe*. Translated and adapted by Jacques P. Cole. London: Harold Starke Publishers Ltd., 1994.

McGuire, Joseph D. *Pipes and Smoking Customs of the American Aborigines, Based on Material in the U.S. National Museum*. Smithsonian Institution. Washington, D.C.: Government Printing Office, 1899.

Mode, Robert L. *Meerschaum Masterpieces: The Premiere Art of Pipes*. Nashville, Tennessee: Museum of Tobacco Art and History, 1990.

Rapaport, Benjamin. *A Complete Guide to Collecting Antique Pipes*. Atglen, Pennsylvania: Schiffer Publishing, 1998.

_____.*Collecting Antique Meerschaum Pipes, Miniature to Majestic Sculpture*. Atglen, Pennsylvania: Schiffer Publishing, 1999.

Robert, Joseph Clarke. *The Story of Tobacco in America*. New York: Alfred A.Knopf, 1949. Reprint. Chapel Hill, North Carolina: University of North Carolina, 1967.

Sudbury, Byron. *Historic Clay Tobacco Pipemakers in the United States of America*. Reprint. Ponca City, Oklahoma: the author, 1979.

West, George A. *Tobacco, Pipes, and Smoking Customs of the American Indians*. Milwaukee, Wisconsin: North American Press, 1934. Reprint. Westport, Connecticut: Greenwood Press, Inc., 1970.

Periodicals

Pipes and Tobaccos.

Pipe Friendly.

Pipesmoke.

Smoke, Cigars, Pipes and Life's other Burning Desires.

The Pipe Smoker's Ephemeris.

SmokeShop.

Pipe Makers Contact Information

E. ANDREW
610 E. Otjen Street
Milwaukee, WI 54207
Telephone: 414-744-5181
Email: eandrew@execpc.com

SHIZUO ARITA
Onoji 4374-7
Machida
Tokyo 195-0064
Japan
Telephone and Fax: +81-42-735-3771
Email: arinko@mtc.biglobe.ne.jp

ASHTON
c/o David Field
935 Caledonia St.
Philadelphia, PA 19128
Telephone: 215-508-9309
Fax: 215-508-5945
Email: Rdfield@voicenet.com

ALFRED BAIER
Butternut Lane
PO Box 2043
Manchester Center, VT 05255-2043
Telephone: 802-362-3371

BANG'S PIBEMAGERI
c/o Uptown Smoke Shop
3900 Hillsboro Road
Nashville, TN 37215
Telephone: 615-292-9576
Fax: 615-263-2822
Website: www.uptowns.com

PAOLO BECKER AND BECKER & MUSICÒ
c/o David Field
935 Caledonia St.
Philadelphia, PA 19128
Telephone: 215-508-9309
Fax: 215-508-5945
Email: Rdfield@voicenet.com

J.M. BOSWELL
J.M. Boswell's Handmade Pipes
586 Lincoln Way East
Chambersburg, PA 17201
Telephone: 717-264-1711

BUTZ-CHOQUIN
c/o Music City Marketing Pipe Source
477 McNally Drive
Nashville, TN 37211
Telephone: 615-292-9576
Fax: 615-263-2822
Website: www.uptowns.com

JOHN CALICH
J. Calich, Pipemaker
3572 Credit Woodlands
Mississauga, Ontario L5C 2K6 Canada
Telephone: 905-277-9192

CASLTEFORD
c/o Music City Marketing Pipe Source
477 McNally Drive
Nashville, TN 37211
Telephone: 615-292-9576
Fax: 615-263-2822
Website: www.uptowns.com

CASTELLO
c/o Castello Pipes North America
PO Box 5179
Woodbridge, VA 22194
Telephone: 703-897-9128
Email: castello@castello.net
Website: www.novelli.it

Jess Chonowitsch
c/o Music City Marketing Pipe Source
477 McNally Drive
Nashville, TN 37211
Telephone: 615-292-9576
Fax: 615-263-2822
Website: www.uptowns.com

J.T. Cooke
J.T. & D. Cooke
1297 Will George Rd.
East Fairfield, VT 05448-9801

Jody Davis
Princeton Pipes
3900 Hillsboro Road
Nashville, TN 37215
Telephone: 615-292-9576
Fax: 615-263-2822
Website: www.uptowns.com

Dunhill
c/o Lane Limited
2280 Mountain Industrial Blvd.
Tucker, GA 30084-5014
Telephone: 770-934-8540
Fax: 770-934-8608

Lee E. Erck
816 W. College
Marquette, MI 49855
Telephone: 906-225-0817
Email: Erck2@aol.com
Website: www.futurehorizons.net

Ron Fairchild
Fairchild Pipes
4302 Dorothy St.
Bellaire, TX 77401
Telephone: 713-667-8892

FERNDOWN
c/o James Norman, Ltd.
260 West 39th St.
16th Floor
New York, NY 10018
Telephone: 800-525-5629

S.M. FRANK & COMPANY, INC.
P.O. Box 789
1000 North Division St.
Peekskill, NY 10566
Telephone: 800-431-2752
Fax: 914-739-3105
Email: smokepipes@aol.com
Website: www.smfrankcoinc.com

HOLGER FRICKERT
c/o C.A.O. International, Inc.
223 Oceola Avenue, Suite B
Nashville, TN 37209
Telephone: 615-352-0587
Fax: 615-353-0610
Email: caointernational@caointernational.com
Website: www.caointernational.com

L.J. GEORGES
c/o Steve Monjure
Monjure International
12 Cedar Creek Drive
Jamestown, NC 27282-8800
Telephone: 336-889-2390
Fax: 336-889-9437

JUN'ICHIRO HIGUCHI
Address and telephone unlisted

IL CEPPO
c/o David Field
935 Caledonia St.
Philadelphia, PA 19128
Telephone: 215-508-9309
Fax: 215-508-5945
Email: Rdfield@voicenet.com

LARS IVARSSON
c/o Uptown Smoke Shop
3900 Hillsboro Road
Nashville, TN 37215
Telephone: 615-292-9576
Fax: 615-263-2822
Website: www.uptowns.com

DAVID JONES
6006 Turtle Creek Dr.
Texarkana, TX 75503
Telephone: 903-838-0854

JØRN
c/o Music City Marketing Pipe Sources
477 McNally Dr.
Nashville, TN 37211
Telephone: 615-292-9576
Fax: 615-263-2822
Website: www.uptowns.com

KARL-HEINZ JOURA
c/o C.A.O. International, Inc.
223 Oceola Avenue, Suite B
Nashville, TN 37209
Telephone: 615-352-0587
Fax: 615-353-0610
Email: caointernational@caointernational.com
Website: www.caointernational.com

ANNE JULIE
Sdr. Kirkerej 8
9940 Laesø
Denmark
Telephone: 98-499504

KIRSTEN PIPE COMPANY, INC.
P.O. Box 70526
Seattle, WA 98107-0526
Telephone: 206-783-0700
FAX: 206-789-5567
Email: Lynn@Kristenpipe.com
Website: www.geocities.com/pipeline/7279/Kristen

W. Ø. LARSEN
9 Amagertorv
DK 1160 Copenhagen
Denmark
Telephone: 33 12 20 50
Fax: 33 15 63 22

SAMUEL M. LEARNED
149 East Market Street
York, PA 17401
Telephone: 717-846-9290
Fax: 717-845-8700

LUCILLE LEDONE
415 Timberline Drive
Appleton, WI 54915
Telephone: 920-993-2997

JAMES MARGROUM
Mr. Groum Pipes
68 Marianne Dr.
Hanover, PA 17331
Telephone: 717-632-6411
Website: Mrgroumpipes@netrax,net

ANDREW MARKS
3641 Route 30
Cornwall, VT 05753
Telephone: 802-462-2112

CLARENCE L. MICKLES
Mickles Pipe Repair
321 Oswego St.
Park Forest, IL 60466
Telephone: 708-748-7293

MASTRO DE PAJA
Attn: Melissa King
6649 Peachtree Industrial Blvd., Ste. M
Norcross, GA 30092
Telephone: 888-3-MASTRO
Fax: 770-449-4904
Email: MastroLTD@aol.com
Website: www.mastrodepaja.com

Elliott Nachwalter
Elliott Nachwalter Pipestudio
Trout Run at River Road
Arlington, VT 05250
Telephone: 802-362-5589
Email: troutrun@sover.net
Website: www.vtpipes.com

Bo Nordh
c/o Uptown Smoke Shop
3900 Hillsboro Road
Nashville, TN 37215
Telephone: 615-292-9576
Fax: 615-263-2822
Website:www.uptowns.com

Nørding
Banegraven 9
3550 Slangerup
Denmark
Email: Nording@post12.tele.dk

Peterson
c/o Hollco-Rohr
20717 Marilla Street
Chatsworth, CA 91211
Telephone: 818-885-0850

Radice
c/o David Field
935 Caledonia St.
Philadelphia, PA 19128
Telephone: 215-508-9309
Fax: 215-508-5945
Email: Rdfield@voicenet.com

Joan Saladich y Garriga
c/o Font Nova, 16
08202 Sabadell
Spain

SER JACOPO
c/o The Marble Arch Ltd.
P.O. Box 966
Rockville Centre, NY 11571
Telephone: 800-The-Arch
Fax: 516-536-9781
Website: www.pipes.tm

MANUEL SHAABI
c/o C.A.O. International, Inc.
223 Oceola Avenue, Suite B
Nashville, TN 37209
Telephone: 615-352-0587
Fax: 615-353-0610
Email:caointernational@caointernational.com
Website: www.caointernational.com

DENNY SOUERS
3280 Braidwood Dr.
Hilliard, OH 43026
Telephone: 614-876-0790
Email: drsouers@aol.com
Website: www.lioncrest.com

STANWELL
c/o Lane Limited
2280 Mountain Industrial Blvd.
Tucker, GA 30084-5014
Telephone: 770-934-8540
Fax: 770-934-8608

TREVER TALBERT
806 Lakeview Dr.
Thomasville, NC 27360
Telephone: 910-476-0318
Email: zoth@hpe.infi.net
Website: www.pipes.org/Trever

William Ashton Taylor
Ashton Pipes
c/o David Field
935 Caledonia St.
Philadelphia, PA 19128
Telephone: 215-508-9309
Fax: 215-508-5945
Email: Rdfield@voicenet.com

Mark Tinsky
American Smoking Pipe Company
HC 88 Box 223
Pocono Lake, PA 18347
Email: mt@AmSmoke.com
Website: www.amsmoke.com

Tsuge Pipe Co., Ltd.
4-3-6 Kotobuki
Taito-Ku
Tokyo 111-0042
Japan
Telephone: +81-3-3845-1221
Fax: +81-3-3845-1225

Julius Vesz
c/o Uptown's Smoke Shop
3900 Hillsboro Road
Nashville, TN 37215
Telephone: 615-292-9576
Fax: 615-263-2822
Website: www.uptowns.com

Roy Roger Webb
2216 Green Acres Circle
Sevierville, TN 37862
Telephone: 423-428-1684

STEVE WEINER
Heathtree Woodcrafts
420 Hunters Way
Fox River Grove, IL 60021
Telephone: 847-516-4274
Fax: 847-516-9607
Email: 76532.620@compuserve.com

TIM WEST
1588 Grayling Court
Columbus, OH 43235
Telephone and fax: 614-761-3465

RANDY WILEY
Wiley Briar Pipes
11611 Tucker Road
Riverview, FL 33569
Telephone: 813-677-8527
Fax: 813-671-9084
Email: Wileypipes@aol.com

Notes

[1] Benjamin Rapaport, *A Complete Guide to Collecting Antique Pipes*, 2d ed. (Atglen, PA: Schiffer Publishing Ltd., 1998), p. 93; Levárdy, pp. 145–6.

[2] Rapaport, pp. 49–50; Levárdy, pp. 118–20.

[3] Rapaport, p. 51 and Benjamin Rapaport, *Collecting Antique Meerschaum Pipes, Miniature to Majestic Sculpture.* (Atglen, Penn.: Schiffer Publishing Ltd., 1999), p. 19.

[4] Ibid., pp. 33–7.

[5] Rick Newcombe, "Breaking in a Pipe," *Pipes and Tobacco.* Summer 1998, pp. 16-7.

[6] Keith Moore, "Julius Vesz, The Grand Old Man of Pipemaking," *Great Carvers of Our Day Series*, vol. 2, video. (Nashville, Tenn.: Uptown's Smoke Shop, 1998).

[7] Ibid., pp. 13-4.

[8] Keith Moore, "The Great Danes," *Great Carvers of Our Day Series*, vol. 1, video. (Nashville, Tenn.: Uptown's Pipe Shop, 1998); Dayton H. Matlick, "Mental 'Model,'" p. 33.

[9] Ibid., pp. 64 and 66.

[10] Moore, "The Great Danes."

[11] Matlick, "Mental 'Model,'" p. 31.

[12] Dayton H. Matlick, "Self-taught Grand Master," *Pipes & Tobacco.* Winter 1996–97, p. 29.

[13] Dayton H. Matlick, "Nature's Designs," *Pipes & Tobacco.* Winter 1996–97, p. 23; Moore, "The Great Danes" video.

[14] Keith Moore, "W.Ø. Larsen: A Family Tradition," *Great Carvers of Our Day Series*, vol. 4, video. (Nashville, Tenn.: Uptown's Smoke Shop, 1999).

[15] Ibid., pp. 21-2; Keith Moore, "St. Claude: The Beginnings of Briar," *Great Carvers of Our Day Series*, vol. 3, video. (Nashville, Tenn.: Uptown's Smoke Shop, 1998).

[16] Matlick, "Butz-Choquin," pp. 21-2.

[17] Ibid., pp. 23-4.

[18] R.C. Hamlin, "The Ashton Pipe Story," www.pipeguy.com/ashtonpi, 1996.

[19] Michael Balfour, *Alfred Dunhill: One Hundred Years and More*. (London: Weidenfeld & Nicolson, 1992), p. 17.

[20] Alan Schwartz, "Dunhill: Behind the White Spot," *Smoke*, Summer 1996, pp. 76-8.

[21] Ibid., p. 79.

[22] Ibid., p. 80.

[23] Alan Schwartz, "The Italian Renaissance," *Pipe Smoke*. Winter 1998–99, pp. 16–8.

[24] Ibid.

[25] Dayton H. Matlick, "Pipe Brebbia," *Pipes and Tobacco*. Fall 1997, p. 44; interview with Steve Monjure, Monjure International, 24 March 1999.

[26] Matlick, "Pipe Brebbia," p. 46; interview with Bob Ysidron, Savinelli, 1 March 1999

[27] Ibid., pp. 49–50.

[28] Schwartz, "The Italian Renaissance," p. 18.

[29] Dayton H. Matlick, "Maestro de Paja," *Pipes and Tobacco*. Winter 1996–97, p. 44.

[30] Schwartz, "The Italian Renaissance," p. 18.

[31] Chuck Stanion and Phil Bowling, "A Pipemaker's Long and Widening Road," *Pipes and Tobacco*. Winter 1999, pp. 41-2.

[32] G.F., "Italian Pipe Production Panorama," *Amici della Pipa*. August 1996, p. 8; Giuseppe Lepri, "Travel Notes," *Amici della Pipa*. August 1996, p. 15.

[33] G.F., "Italian Pipe Production," p. 8.

[34] Jan Andersson, "Bo Nordh," *The Pipe Smoker's Ephemeris*. Summer 1994–Spring 1995, pp. 19-21.

[35] Stanion & Bowling, "A Pipemaker's Long and Widening Road," pp. 40-4.

[36] David Jones, "David Jones in His Own Words," *The Pipe Rack*. October 1996, pp. 10-1.

[37] Ibid.

[38] Ibid., p. 57.

Photography Credits

Courtesy of Alfred Baier: p. 134, p. 135, p. 138; Photo by Arita: p. 121; Photo by Erik Borg: p. 164; Photo by J.M. Boswell: p. 141; Photos by Jun'ichiro Higuchi: p. 58, p. 122 ; Photo by David Jones: p. 154; Courtesy of Lioncrest, Inc.: p. 171, p. 181, p. 183, p. 184, p. 187; Photo by Elliot Nachwalter: p. 169; Courtesy of Benjamin Rapaport Collection: p. 18, p. 27, p. 28; Courtesy of Joan Saladich y Garriga: p. 32, p. 41, p. 43, p. 126; Courtesy of Mark Tinsky: p. 176; Courtesy of UST, Inc.: half-title page, p. 10, p. 12, p. 15, p. 20, p. 21, p. 22, p. 25, p. 26, p. 30, p. 31, p. 35, p. 36, p. 37; Courtesy of Roy Roger Webb: p. 56, p. 179.